5
D

CHINESE BRONZES

Christian Deydier

CHINESE BRONZES

Foreword by Michel Beurdeley

Translated by Janet Seligman

Rizzoli
NEW YORK

Translated from the French:
Les Bronzes chinois

Copyright © 1980 Office du Livre S.A.,
Fribourg, Switzerland

English edition published 1980
in the United States of America by:

*R*IZZOLI INTERNATIONAL PUBLICATIONS, INC.
712 Fifth Avenue/New York 10019

Library of Congress Catalog Card Number: 80-51170
ISBN 0-8478-0323-6

Printed in Switzerland

Contents

Foreword

by Michel Beurdeley

Taotie Mask

horn   nasal rib   horn   tail

upper jaw   eye   claw   paw

division

kui dragon

In asking me to introduce this book I imagine that you are addressing the collector of ancient bronzes rather than the specialist in Chinese art, and so I shall try to write in that spirit about my personal experiences, my joys and disappointments. It will be a difficult task for I have trodden many paths in the world of art, traversing both time and space! To quote the words of a celebrated antiquarian and a very dear friend — now no longer with us: 'One is born an art-lover, one becomes a collector'. The sequence of events is inescapable. At the outset all seemed simple: besides a smattering of knowledge, all that was needed was a little taste, a little flair, method and money. And when I say 'a little', obviously 'much' would enable one to broaden one's horizon considerably. The three elements of success as defined by Balzac — 'the legs of a stag, the time of an idler and the patience of an Israelite' — belong to the realm of literature rather than to present-day artistic life.

An ancient bronze must possess certain intrinsic qualities to command attention at first glance: the volumes must be well balanced, the material dense, form and decoration must harmonize, design must be clear-cut, sculpture bold but not heavy. However complex the ornament, it must give an impression of coherence and symmetry. One must also ascertain that the object is in good condition: cracks, marks of wear and dents all impair its quality. If pressed, one may admit, with André Malraux, that 'some damage, which might be the work of a talented antique-dealer, has style'.

The patina which results from long years of burial in the ground is another feature which the collector regards as indispensable. Patina has always been appreciated in China, contrary to the opinion of certain writers who base their judgment on the imperial bronzes in the Taiwan museum, the surfaces of which have been wax-polished, giving them a dark and not very attractive tonality. An often quoted anecdote shows that this enthusiasm for patina dates back to very early times: in the year 113 B.C. a beautiful sorceress unearthed a tripod *ding [ting]* which, though lacking an inscription, had a patina that was 'brilliant, unctuous and iridescent like a dragon'. Emperor Wu considered the discovery so important that he decided to name the era that began with his reign the 'era of the tripod'. Much later, under the Ming, the collector Zhang Ying-wen [Chang Ying-wen] studied the various aspects of bronzes and specially admired those with a fine emerald-green patina. His enthusiasm bordered on pathological delirium. He held that in certain circumstances — in rainy weather, by lamplight, when drunk or in a mercenary spirit — an art-lover should abstain from collecting and, as a crowning act of ostracism, he excluded women from the joys of collecting.

Having considered the various qualities of a bronze, one must determine its date by reference to its stylistic features. This is not too difficult nowadays, as I am sure art-lovers will agree when they have read this book. It is also interesting to be able to read the inscriptions that occur on certain bronzes: 'good wishes for your anniversary', 'hopes for a peaceful life', 'entreaties to the spirits for a limitless number of children'. Of the 12,000 bronzes recorded under the Song [Sung] dynasty, a third had engraved inscriptions — which, according to Ying wei, is an unassailable criterion of quality: 'However old it may be, a bronze without [an] inscription has no history. It raises hypotheses but offers no certainty'. To quote Ying wei again, 'certain aural and olfactory tests should enable one to discover the age of a bronze; when struck an ancient bronze emits a clear note'. But knowledge is sometimes best forgotten; in art, as in love, instinct is the surest guide.

It may appear indecent for an art book to raise the subject of speculation, of filthy lucre, but it has in fact existed at all periods. As early as the Shang and Zhou periods bronzes were identified as part of a prince's treasure *(zheng bao [cheng pao])*; they were looked upon as precious objects in which the spirit of the Chinese people was expressed in all its complexity. The process of casting a bronze had some of the quality of magic: while young virgins worked the bellows, the founder and his wife threw themselves into the furnace, thus accomplishing the ritual sacrifice. Naturally such pieces were sought after with as much frenzy as Greek vases in the Renaissance. It was not uncommon during a siege to induce the attackers to retreat by offering them objects made of bronze. And then there was the official who pardoned an accused man who offered him an ancient bronze.

From the Song period onwards, bronzes, now stripped of their good and evil powers, increasingly became collectors' items and were sought not only for the delectation of art-lovers but also because there was always a religious respect for antiquity. So the prices of bronzes will come as no surprise: they have always been high, but today they reach astronomical figures. In March 1977 a *gui [kuei]* from the former Malcolm Collection, dated 1010 B.C. was sold by Sotheby Parke Bernet for £110,000. In the following year in Paris, a large *zun [tsun]* from the end of the Shang dynasty, decorated with three *taotie [t'ao-t'ieh]* masks and with a fine green patina partly encrusted with traces of malachite and lapis-lazuli, fetched a similar sum. Works of art must surely be the playthings of those who aspire to be considered the great ones of this world. As Oscar Wilde wrote in *The Portrait of Dorian Gray:* 'Beauty... makes princes of those who have it.'

My own entrance into this field was far more prosaic. In 1952, the Marquis de Ganay commissioned me to put his collection of ancient bronzes up for auction. This was a heavy responsibility at a difficult period when private collectors were rare, dealers reticent and museums without money. I had decided to bid to keep prices up — as far as my 'savings' would allow — but in the event everything sold for far more than I had expected. Robert Rousset, whom I had asked to act for me in New York, obtained some sizeable commissions from a collector in Chicago. Jeanine Loo bought back the bronzes inlaid with gold and silver that her father had sold thirty years earlier to the Comtesse de Béhague; the Musée Guimet preempted a pair of monkeys — bronze feet inlaid with silver, from a low table — of which the other pair were in Kansas City and the little recumbent buffalo from the former Wannieck Collection. Which meant that there was nothing left for me. At this juncture I left for the Far East and bought from T.Y. King (the great Shanghai dealer, who had been obliged to retreat to Hong Kong) a very fine *ding* of the Shang period. I then went to Japan, where I visited the firm of Kawai and bought a few bronzes — but I was rather late and the finest pieces had been reserved by Vadime Elisseeff for the Musée Cernuschi in Paris. Shortly afterwards, while I was in Japan, my friend Sammy Lee took me to the outskirts of Kyoto, where an industrialist collector was in financial straits. We stayed in a small Japanese hotel; at nightfall, the bronzes were brought to us with the usual ceremony. They had been carefully wrapped in purple silk and were lovingly removed from great wooden boxes. Despite the dim light, we at once distinguished three exceptional pieces: a large *pan* with little birds on the rim, rather like the one in the Hakutsuru Museum in Kobe, two little recumbent buffalos, the larger of which had been in the Alfred F. Pillsbury Collection in the United States, and a marvellous *hu* with a jade-green patina and an inscription inside the cover: 'Hiang has had this precious vessel made for his dead father; may his progeny be brilliant!' The three bronzes could be dated to about the eighth century B.C. For the period, the price was startling. But did not Wilde say: 'The only way to get rid of a temptation is to yield to it'? Which we did, unhesitatingly.

The great sales of Chinese art in London were a privileged meeting-place for collectors from all over the world. These were collectors for whom intuition was not mingled with modishness, nor passion with snobbishness, nor enjoyment with speculation. There, like pieces in a kaleidoscope, was one of the Bluett brothers with John Sparks (a fellow-dealer and also a friend) with a carnation in his button-hole and his fine white hair in unruly waves. They were seated one to the right, the other to the left of the auctioneer, discreetly pushing up the bidding, each having commissions in his pocket from Americans: Vanderbilt, Rockefeller and Avery Brundage, the future president of the Olympic Games; Frenchmen: Michel Calmann and David-Weill and Englishmen: Denis Cohen, Lord Cunliffe, Sir Alan Barlow or C.G. and B.Z. Seligman. The elegant Tai of New York or my dear friend Edward Chow of Hong Kong, would sometimes upset this orderly play with their determined bidding.

In 1958, as soon as the Chinese frontiers — known jokingly at the time as 'the bamboo curtain' — were opened to tourists, Maurice Rheims and I decided to go to Peking. How our imaginations ran riot! There would be stalls overflowing with marvels from the Ancien Régime, rejected by the new society, antique-dealers anxious to tempt us and accommodating officials. Alas, we soon had to change our tune. Like Bao-yu, the hero of *The Dream of the Red Chamber*, who was distressed to discover that he was unable to find a single object worthy of his covetousness, I found nothing that I liked — or at least nothing that could be bought, for it was forbidden to purchase any object over one hundred years old; therefore, only the sad sin of envy was permitted us. I could not help thinking of those happy days around 1935 when French and other foreign collectors freely purchased marvellous things unearthed by the pickaxes of the tomb-robbers in the Anyang district. Henan [Honan] province in those days teemed with archaeological treasures, and the plunderers, barely heeding stratigraphical layers, bulldozed whole pits and brought out a jumble of jades, inscribed bones, bronzes and pottery. Indeed, they behaved like true bandits, were armed with guns and had no hesitation in smashing and scattering skeletons to escape the vengeance of the 'spirits' whom they feared much more than the official archaeologists of the Academia Sinica. In the quarters of Peking named Liu li chang, Dong su bai, Qian men and Xing long dian and in the Dong an shi market, stalls overflowed then with authentic bronzes that would be a credit to the finest collection today.

The occupation of China by the Japanese in 1937 put an end to official archaeological campaigns. Clandestine excavation continued, but it was no longer the Western camp that benefited from the finds. Not until after 1949 did the archaeologists of the People's Republic initiate genuinely scientific excavation campaigns at other sites. They reaped a harvest of discoveries, some of which were seen in the exhibition *The Genius of China* in London in the winters of 1973 and 1974. Further advances have been made, and the travelling exhibition *Treasures from the Bronze Age in China* that was at the Metropolitan Museum of Art in New York in 1980 is even more astonishing. On display were five hundred objects, mostly bronzes of the Shang and Zhou periods, including a *zu* wine vessel in the form of an owl from the tomb of Princess Hao, wife of one of the last Shang kings. This extraordinary group has turned the old assumptions about the geography of Chinese archaeology upside-down. Now research must not be directed only towards the ancient Shang capital, Anyang in Henan province; other peripheral provinces—Hebei [Hopei], Shanxi [Shansi] and Gansu [Kansu]—also deserve to be explored.

Another fact has emerged from the New York exhibition: it has proved that a realistic style existed before the Eastern Han (206 B.C.–A.D. 24). Until now scholars believed that the art of the earlier periods was hieratic in the main and based on abstraction, though numerous zoomorphic elements were always stylized and were closer to magical symbols than to the animals concerned.

Now vessels have been found decorated with half-figures of rams, tigers, a crane or a rhinoceros that are worthy of the best animal artists of any culture.

In conclusion, let us thank Christian Deydier, one of the few Frenchmen who is able to decipher ancient Chinese characters, for giving us this book with its remarkable text and illustrations. It summarizes all the archaeological information that has been gathered during recent years, and the time is ripe for it.

For myself, as I recall fifty years of the 'unpardonable industry' that is collecting, I find I have suddenly become a spectator of my own life. But, as is well known, the drama of old age is to remain young:

> I do not like the bronze mirror
> For it is too bright,
> It reflects my white hair.

But let us take heart! Those lines by a Tang poet cannot apply to a collector, who, by definition, is always full of enthusiasm, projects and desires. He is in no danger of ever seeing grow up between him and the artifact he cherishes a pact of indifference such as Hemingway observed between the fish and the angler on the banks of the Seine.

Michel Beurdeley
Paris, June 1980

Introduction

The Transcription of Chinese

The Chinese have worked out a new system of transcribing their language into the Roman alphabet. This is called 'pinyin'. It gives a rendering of standard spoken Chinese, 'putonghua' (*p'u-t' ung-hua*). Many Western universities have abandoned their traditional systems of transcription for 'pinyin', which has the advantage of being international. In 'pinyin' polysyllabic expressions are written as one, whereas in the older systems they were separated by hyphens, except for the names of provinces. Since 'pinyin' is becoming increasingly widespread, we have only used the older Wade-Giles transcription in square brackets at its first occurrence, for the sake of the many readers unversed in the new method.

Ever since I decided to write this book, people have asked why I chose to write about Chinese bronzes rather than Chinese pottery, sculpture or jades. I see many possible replies.

The first takes the Chinese viewpoint as its basis. From earliest antiquity, i.e. from the sixteenth or fifteenth century B.C., the Chinese looked upon bronze as the noblest and most venerable substance. It was evidence of power and of riches, and it possessed a supernatural and religious quality. Later, from the fourth or fifth century A.D., the religious aspect, arising out of the introduction of Buddhism into China, continued to maintain bronze in its place among the noble substances.

The second reply assumes a Western point of view. Chinese bronzes, and more particularly the ancient vessels from the end of the Shang and beginning of the Zhou dynasties, are considered by connoisseurs to be the major art of ancient China. For these purists the high point of this art occurred in the twelfth to eleventh centuries B.C.

The third reply is much more down to earth, for it recognizes the fact that we have no introduction to Chinese bronzes from the beginning to the Ming period. At present, the art-lover in search of a general conspectus of the history of Chinese bronzes has to extract his information from many different books in English, French and particularly in Chinese and Japanese.

Yet the question 'why write about bronzes?' can also find a ready answer in the explanation of 'why these objects were produced'. Leaving out of account the Song, Yuan and Ming pieces, which art-lovers consider to be archaizing and over-decorated, Chinese bronzes can be divided into two distinct groups: a) ancient bronzes made between the fifteenth and the third centuries B.C.; b) Buddhist bronzes, made mostly between the fourth and ninth centuries A.D.

The uses to which the bronzes in each of these categorie s were put were also different. The ancient bronzes were reserved for kings, emperors and the great ones of the realm: the common people were excluded. These appurtenances of the élite embodied religion in terms of the rites of ancestor-worship, religious as well as spiritual power acquired directly from 'Heaven' or from the *Shangdi [Shang-ti]* (High Sovereign), coupled with a certain artistic brilliance if they were buried with the dead. More rarely, especially under the Zhou [Chou] dynasty, the casting of a vessel might commemorate an important event, such as a military victory, an act of bravery or a royal favour. Furthermore, the quality of the casting and the refinement of the ornament and work-manship would seem to vary according to the owner's rank. Indeed, the quality of Shang and Zhou bronzes is very unequal. Some pieces are sumptuous, but many are mere receptacles of no artistic or iconographic interest. Scientific excavation would appear to indicate and substantiate clear-cut class divisions in these primitive societies, and these are reflected in the considerable differences in the casting and the power of the ornament in ritual vessels made for ancestor worship.

Buddhist bronzes represent the second period of the art of bronze-founding in China. They were made exclusively for religious purposes and testify to the religious ardour of a people given over to a foreign belief. These bronzes were made for all social classes, and anyone could commission a founder to make a statue for his family shrine. These pieces, some of which are of great artistic merit, mark a total break with the art of bronze of the earlier periods. But they brought about a revival of the art of bronze, which had been declining rapidly since the middle of the Zhou dynasty and during the Period of the Warring States.

Despite many theories, none of them convincing, the origin of bronze-founding in the Chinese world remains enigmatic. Although there is as yet no explanation of the appearance of this technique, one thing is certain and has been proven by excavations at the sites of Zhengzhou [Cheng-chou] and Panlongchen [P'an-lung-ch'en]: bronze was being used towards the middle of the Shang dynasty (*c.* sixteenth-fifteenth century B.C.) to make ritual vessels and weapons.

Yet the scientific evidence relating to the bronzes unearthed at Zhengzhou, which has been corroborated by many excavations undertaken by Chinese archaeologists, has been contested. William Watson considers Zhengzhou to have been a religious centre contemporary with Xiaotun [Hsiao-t'un], the Anyang site, and states that 'it is no longer so certain that the Zhengzhou style' preceded that of Anyang. He goes so far as to add that the Zhengzhou style is one of 'regional conservatism'. This rebuttal of the discoveries derived from the scientific excavations of recent years would seem to take us back some thirty years ago, to the days of the pioneers, when the existence of Zhengzhou, and presumably of the capital Shang Ao, was no more than a literary reference and was viewed with scepticism. Watson adduces no proof to support his theory apart from an explanation apparently connected with the encircling wall and the shape of the city. This provides him with a pretext for stating that Xiaotun/Anyang did not receive the status of royal capital and that its role was purely religious, because it is situated close to the royal graves. If we extrapolate from Watson's assertions, we are faced with the conclusion that Xiaotun was the religious capital of the Shang while Zhengzhou was their royal capital. Further, since these two cities were contemporary at a certain period, it would be reasonable to conclude that the Zhengzhou and Anyang bronzes were contemporary. Prudently, Watson does not draw this conclusion, since all the scientific evidence and the literary and historical texts prove the contrary. One thing is practically certain: Zhengzhou was founded at the latest in the seventeenth century B.C., while Anyang was founded in about the fourteenth to thirteenth centuries B.C. Since the top cultural stratum at Zhengzhou is contemporary with Anyang, it is probable, even certain—as the R 2012 and R 2017 *gu* unearthed at Anyang would seem to prove—that pieces of similar type, i.e. of the style known as Zhengzhou, were made in about the fourteenth to thirteenth centuries at both Zhengzhou and Anyang.

These differences of opinion over the earliest Chinese bronzes need not detain us, for the history of the art of bronze is relatively simple and changes with the variations in religious practice.

During the Shang period, an extremely complex and highly organized from of ancestor-worship governed the whole working of the state. While bronze weapons were the exclusive property of the élite and of the ruling class, the purpose of bronze vessels was purely religious or ritualistic. As the inscriptions would seem to indicate, the latter were usually made to honour the memory of an ancestor. (The short Shang inscriptions are of the type 'made for Father X', 'precious sacrificial vessel made for so and so', etc.) With the advent of the Zhou dynasty, the purely religious purpose of bronze vessels seems to diminish, and their role becomes much more commemorative or honorific. Thus some Zhou inscriptions are of the type 'so and so has had this precious sacrificial vessel made for so and so'; others commemorate a journey, a military campaign, a hunting expedition, an official ceremony or some other religious or secular event in the kingdom.

During the Periods of the Spring and Autumn Annals and the Warring States, the use of bronze spread to wider sections of society. Rich merchants and landowners commissioned bronze vessels from founders, but the vessels' role became exclusively funerary. Production increased considerably, and since the original purpose of the vessels had been forgotten, the quality of the decoration deteriorated. The art of bronze entered upon a period of decline. This became more noticeable under the Han dynasty, when only those objects that were made for princes and emperors retained their sumptuous character thanks to the fact that they were inlaid and gilded—a survival of the wealth and ostentation of that period. A minor and short-lived revival of the art of bronze accompanied the animal style that reflected the influence of the art of the steppes. The decline in the art of bronze in China would appear to be due to the disappearance of ancestor-worship, the spread of Confucianism and Taoism, the cessation of human sacrifices and the introduction of *mingqi [ming-ch'i]* tomb figures.

In the fourth century A.D., and subsequently under the Wei and the Northern Qi [Ch'i], the introduction and expansion of Buddhism in China was reflected in a renaissance or revival of the art of bronze. Ritual vessels were abandoned, and this precious and venerable material was used to make gilded steles. These objects were purely religious in purpose and depicted the various Buddhas, bodhisattvas and other divinities of the Buddhist pantheon, very often under the influence of Indian art. This religious art continued under the Sui and the Tang. But the anti-Buddhist reaction, initiated by the Taoists and Confucianists, culminated in 845 in the proscription of Buddhism. This brought with it a new decline in the art of bronze-founding. However, with the weakening of the influence of Greco-Buddhist Gandharan art, figures of Guanyin were made, and the influence of Persia, now growing as a result of many commercial exchanges, gave rise to vessels of new shapes.

Under the Song dynasty, the charm of the past, archaeological research and the study of the classical texts brought about a revival of ancient traditions and of the art of bronze-founding. This was reflected in the creation of vessels in the style of the ancients but with clear reminders of contemporary taste.

Under the Yuan, the influences of Mongolian art and Buddhism were strong, but there was a minor revival of the art of bronze with the creation of small Buddha figures and bodhisattvas.

The Ming returned to the spirit of the Song, which accounts for the great difficulty of dating works from these two periods and the brevity of the chapter on the Song, Yuan and Ming. Nevertheless, the Ming occasionally created shapes that differed slightly from or were interpretations of the earlier shapes. Buddhist art under the Ming was powerfully influenced by Tibetan art; the best pieces date from the reign of the Emperor Yongle. They are strongly Tantric and are Sino-Tibetan.

The Shang Dynasty

Chronology of the Shang Dynasty

(Dates according to Dong Zuobin [Tung Tso-pin])

Non-dateable Kings

Da Yi [Ta-i] or Tang [T'ang]	Xia Xin [Hsia Hsin]	1370 B.C.
Da Ding [Ta Ting]		
Da Jia [Ta Chia]	Xiao Yi [Hsiao I]	1349 B.C.
Wai Bing [Wai Ping]	Wu Ding [Wu Ting]	1339 B.C.
Da Geng [Ta Keng]	Zu Ji [Tsu Chi] or Xiao Ji [Hsiao Chi]	[?]
Xiao Jia [Hsiao Chia]	(Dong Zuobin doubts whether he reigned)	
Da Wu [Ta Wu]		
Yong Ji [Yung Chi]	Zu Geng [Tsu Keng]	1280 B.C.
Zhong Ding [Chung Ting]	Zu Jia [Tsu Chia]	1273 B.C.
Wai Ren [Wai Jen]		
Jian Jia [Chien Chia] or Hedan Jia [Ho-tan Chia]	Lin Xin [Lin Hsin]	1240 B.C.
Zu Yi [Tsu I]	Kang Ding [K'ang Ting]	1234 B.C.
Zu Xin [Tsu Hsin]		
Qiang Jia [Ch'iang Chia] or Wo Jia [Wo Chia]	Wu Yi [Wu I]	1226 B.C.
Zu Ding [Tsu Ting]		
Nan Geng [Nan Keng]	Wen Wu Ding [Wen Wu Ting]	1222 B.C.
Hu Jia [Hu Chia] or Yang Jia [Yang Chia]	Di Yi [Ti I]	1209 B.C.
Pang Geng [P'ang Keng]	Di Xin [Ti Hsin]	1174-11 B.C.

The appearance at the end of the last century in many antique dealers' shops of fragments of bone and tortoise-shells, covered with inscriptions prompted a renewal of interest in research into the Shang dynasty. The main aim of this research was to establish the provenance of these objects. It led to the discovery and excavation of many ancient remains in a place called Yinxu [Yin-hsü], or the 'Waste of Yin', as Menzies was to call it. These campaigns, which were private at first, but led by the Academia Sinica after 1928, proved that these were the remains of a large Shang town of the ancient period. The town turned out to be the last capital of the Shang dynasty. Situated near the town of Anyang or, more precisely, at the site of the village of Xiaotun [Hsiao-t'un], the site is now celebrated for the richness, quality and quantity of the objects excavated in the course of fifteen scientific campaigns mounted between 1928 and 1938, and then annually from 1950 onwards.

Since the setting up of the People's Republic of China, research on the Shang sites has been extended to all the provinces of the country. The result has been impressive: over three thousand burial sites have been excavated at Zhengzhou, Luoyang, Huixian [Hui-hsien] in Henan [Honan] province, Qufu [Ch'ü-fu] in Shantung province, Quyang [Ch'ü-yang] in Hebei [Hopei] province, etc.

Before embarking on an examination of Shang bronzes, it may be useful to summarize what is known of the society and religion of that ancient era.

POLITICAL STRUCTURE

Knowledge of Shang society is derived mainly from the oracular texts inscribed on bones and the shells of tortoises. A detailed study of this ancient society would be tedious and altogether too vast for the present book. So a rapid summary of the main elements will be given. The reader is strongly advised to refer to the excellent and monumental study in French by L. Vandermeersch entitled *Wang Dao ou la Voie Royale*.

Baronial Families

Both Ding Shan and Shirakawa Shizuka call them clans; Vandermeersch terms them territorial families. They were the 'estates' surrounding the royal domain proper, and so they bore the name of their founder and conqueror. The baronial families paid special dues in the form of tortoise-shells and shoulder-blade bones, for the royal court used enormous numbers of these in the pursuance of the rites of divination. The texts in which these shells and bones are mentioned form a special category known as 'apophysial inscriptions' and consist of phrases of the following type: 'on such a day such a one [name of person] brought a tribute of [a quantity of bones or shells,

1
Jue, flat base with a band of *taotie* masks on the body.
Bronze with dark green patina
Shang dynasty, Zhengzhou period, end of sixteenth-beginning of fifteenth century B.C.
Musées Royaux d'Art et d'Histoire, Brussels

1

2

2
Fang ding with decoration of *taotie* masks and raised dots. The *taotie* masks on the legs are supplemented by stylized cicada wings.
Bronze with green patina

Shang dynasty, end of the Zhengzhou period, fourteenth century B.C.
Height 22 cm.
Private collection: sold C. Boisgirard and A. de Heeckeren, Paris, 1980

or even of prisoners]. Signed by [name of the annalist who had recorded the event]'. The tribute was given by princes, high-born ladies or toponyms (baronial families founded by eminent officials), many of which are known to have been races of foreigners, tributaries to the Shang state.

The 'Zi' [Tsu]: Sons and Princes

The *jiaguwen [chia ku-wen]* inscriptions mention many zi, who fall into two groups:

a) those who bore a cyclical name: Zi Ding [Tsu Ting], Zi Geng [Tsu Keng], Zi Gui [Tsu Kuei], etc. In this particular case, zi may be translated by 'son' because it denotes membership in the royal family. If the king of their generation sacrificed to them, he called them *xiong [hsiung]*, meaning 'elder brother'; the king of the next generation would call them fu = 'father' (in the sense of father-uncle). These two elements corroborate the translation of zi as 'son', for in these particular cases the zi are recognized as a king's own sons.

b) those whom the inscriptions call zi, but under their personal names. These were the princes or sons of the kingdom, resident in the capital and belonging to the *duozizu [to tzu-tsu]*, the corps of princes. They took part in military operations and in the restoration of order at home. They accompanied the king on hunting expeditions and sometimes replaced him at sacrifices. These princes would appear to have been the most highly placed members of the Shang community.

Ladies: *Fu-X*

The expression *fu-x* either denotes a name borne by the wives of princes, or, more probably, it was a feudal title given solely to women. Whatever their status, these ladies performed important functions, just like the princes. They are found on military operations, presenting the tribute of their native baronial family and personally making certain that it is handed over. They played a part in agricultural affairs and occasionally even replaced the king at the ritual ceremonies. This meant that the activity of these ladies was important politically, mainly in the role of mediators between the capital and the baronial families.

3

3
Lihe with bulbous legs and a handle at the side. Decoration of stylized *taotie* masks in low relief on the body.
Bronze with light green patina
Shang dynasty, Zhengzhou period, end of sixteenth-beginning of fifteenth century B.C.
Height 23.5 cm.
Asian Art Museum, San Francisco, Avery Brundage Collection

Vassals and Pages

In the Shang aristocracy, which must be considered an aristocracy of rank, there were two categories below the rank of prince: the *chen [ch'en]* or *duochen [to-ch'en]* and the *xiaochen [hsiao-ch'en]*, that is, vassal and page respectively. The vassals were the subjects nearest in rank to the princes, while the pages represented the rank immediately below them.

The People or *Zhongren [Chung-jen]*

The term *zhongren,* by which was long understood the slaves who laboured in the fields, is found most often in inscriptions relating to war. The phrases are of the type: 'will so-and-so be ordered to take the mass [or the people] and to attack such and such a person [or such and such a foreign race]'? The 'mass' *zhong* was also the object of a certain usage *zai [tsai],* on the occasion of which the ceremony *gao[kao]* to the ancestors was celebrated. These various points convince Vandermeersch that *zhong* denotes the 'mass' of the Shang race, because 'it is called upon not only to work in the fields but also to make war, and the king is deeply worried at the losses that it might sustain, which appears to be the purpose of the summons to religious assemblies'. This 'mass' representing the bulk of the community may be considered to be the 'mass' of the king's people.

SOCIAL STRUCTURE

The *Yin* or Ministers

The word *yin,* head of an administrative department, may be translated by 'minister' or 'royal counsellor'. It was probably the title that was reserved for the king's brothers or for princes, nephews of the king, who had been raised to this highest rank of the Shang aristocracy.

The *Ya* or Wardens

This office ranked immediately below the *yin* (ministers) in the hierarchy. In the oracle inscriptions, the *ya* are expected to perform the ceremonies of the *yu [yü]* type which were offered to obtain the protection of the spirits. They are also entrusted with military expeditions and with moving the population.

Yet despite their military activities, from the shape of the character *ya,* which suggests the royal graves, Shirakawa Shizuka believes that they were the officials of the most important tombs and therefore essentially royal.

Soldiers

This category embraces the *ma* (marshals), *quan [chüan]* (huntsmen), *she* (archers), *shu* (halberdiers) and *pang [p'ang]* (guards). It would appear from the *jiaguwen* that the archers formed the major part of the armed forces and were organized into three corps of 125 men each. The halberdiers made up five corps. The marshals represented the shock-troops of the Shang army; they were the combat forces in chariots. Each division was composed of twenty-five chariots and was sub-divided into five groups of five chariots. On military campaigns, and sometimes on hunting expeditions, the royal armies consisted of between three and five thousand armed men.

Feudal Titles

The Hou *(Marquises)*

According to Vandermeersch the marquises were archers who had distinguished themselves hunting. As he says: 'hunting expeditions took place in every part of the country and were naturally attended by the members of the Shang community settled in the region. The title of distinguished archer (i.e. marquis) must have been given on such occasions to the heads of the baronial families in those places which had proved to have the best archers. These marquises did not receive fiefs but were probably given command of a fortified town in these extra-metropolitan areas.'

The Bo [Po] *(Counts)*

Chen Mengjia believes that there were two kinds of count: the *bo* and the *fangbo*. The former, counts of the Shang country, were high officials of the kingdom who had distinguished themselves by their qualities. Such is the case of the famous general Zhi Ji [Chih Chi], who received this title in recognition of his victories against the barbarian tributaries. The *fangbo* were foreign rulers who had submitted to, and subsequently been integrated into, the Shang state.

4
Ding with *taotie* masks on the upper part and the legs.
Bronze with a green patina
Shang dynasty, end of the Zhengzhou period, fifteenth century B.C.
Height 35.5 cm; diameter 28.5 cm.
Tai Collection, New York

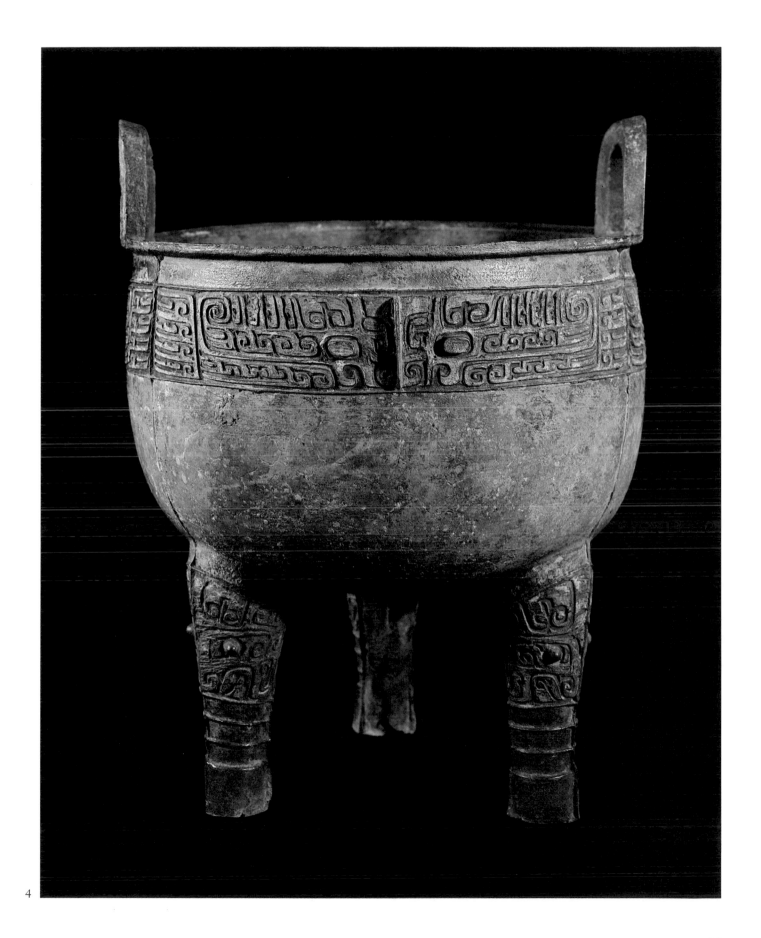

4

5

RELIGION

Most of what is known of the religion of the Shang is derived from a study of the *jiaguwen* inscriptions. This cult, which does not in any way correspond to what we learn from the classical texts, can be divided into two entirely distinct parts.

Cult of the Divinities

The king, who was the liturgical leader, worshipped certain deified phenomena of the natural world. Sacrifices were always preceded by divination with the aid of bones or tortoise-shells and were offered in the hope of receiving either rain or good harvests. Considerable ceremonies were conducted to this end, in the course of which animals—oxen, rams, pigs and dogs—and sometimes human beings were sacrificed. In some cases a single creature was sacrificed; at other times as many as a thousand animals were offered. They were killed as sacrifice, quartered, burnt, buried alive or drowned in a river. The humans, exclusively Qiang [Ch'iang] prisoners (a race that was perpetually at war with the Shang), were beheaded with a halberd.

This immoderate sacrifice was made principally to the Yellow River, called *he [ho]* in the oracle inscriptions. But other phenomena of the natural world were the object of a similar cult, though it was less frequent and on a smaller scale. They were: rivers: *huan* (or *yuan*) *[yüan]*, *shang*, *you [yu]*; mountains: *yue [yüeh]*, *xiong [hsiung]*, the Five Mountains, the Jade Mountain; the earth *tu [t'u]* and the local lands; the cardinal points; and the *di [ti]* or 'Sovereign on high.' He represented the superior divinity of the hierarchy who commanded the natural and supernatural elements and the destiny of men. Requests were not addressed to him directly. A typical oracle text referring to this supreme deity would run: 'Will the Sovereign on high command rain'? Sacrifices were made to his representatives, mainly the winds.

6

5

Zun or *lei,* shouldered. Band of *kui* dragons and *taotie* masks on neck and body.
Bronze with green patina
Shang dynasty, Zhengzhou period, fifteenth-fourteenth century B.C.
Height 26 cm.; diameter 20 cm.
Arthur M. Sackler Collection, New York

6
Jia with two bands of stylized *taotie* masks.
Bronze
Shang dynasty, Zhengzhou period, fifteenth-fourteenth century B.C.
Height 33.5 cm.
Asian Art Museum, San Francisco

Contrary to Granet's assertion, it is interesting to note that the *jiaguwen* contain no reference to a cult of the sun or moon.

Ancestor Worship

Side by side with the incoherent cult addressed to the deities went a highly organized and cyclical ritual performed in memory of the royal ancestors. Ancestor worship was central to Shang religion and was expressed in an extremely elaborate and complex liturgy.

Two forms of the ceremony are recorded in the oracle inscriptions:

a) the question may turn either on whether or not to perform a ceremony or in what manner it should be performed. 'On the day *wuyin*, divination by Bing [Ping]. Question: should Fujing [Fu-ching] (Lady Jing) conduct the ceremony *yu* [in honour of] the mother, *geng [keng]*? 'Divination on the day *jiashen [chia-shen]*. Should Lady Shu conduct the ceremony *yu* [in honour of] Bi Ji [Pi Chi] [with the sacrifice of] two cows? Should Lady Shu conduct the ceremony *yu* [in honour of] Bi Ji [with the sacrifice of] an ox and a sheep'?

b) the divination may refer to something other than a sacrifice, as in the case of inscriptions relating to future decades. Sometimes a certain sacrifice that is to be made during the coming decade is mentioned following the question: 'On the day *guiwei [kuei-wei]* divination by

Yong [Yung]. Question: the king [asks] will the coming decade be without ills? In this first month, on the day *jiashen* the sacrifice *ji [chi]* will be made to Zu Jia [Tsu Chia]; the sacrifice *xie [hsieh]* to X-Jia [X-Chia].' 'On the day *guimao [kuei-mao]*, divination by the king. Question: will the next decade be without ills? In this fourth month, the king has given the omen in the words: very favourable. On the day *jiashen* the sacrifice *rong [jung]* [will be made] to Da Jia [Ta Chia].'

The findings of current research on ancestor worship would seem to suggest that the liturgy was based on the celebration of five sacrifices in turn: *rong [jung]*, *yi [i]*, *ji [chi]*, *za [tsa]* and *xie [hsieh]*; the repetition of these made up a regular cycle.

THE ARCHAEOLOGY AND ART OF BRONZE

The Site of Zhengzhou

In the opinion of some scholars, the site of Zhengzhou, discovered in 1950 in Henan province, represents the 'city of Ao', capital of the kingdom under Zhong Ding, the tenth Shang king. The town was rectangular in shape and measured 2 kilometres from north to south by 1.7 kilometres from east to west; like the Neolithic settlements of the Longshan culture it was surrounded by a wall of rammed earth, *hangtu [hang-t'u]* (height: 5 m,

7

thickness at base: 20 m). Excavation is at present in progress at the Renmin gongyuan [Jen-min kungyüan], Erligang [Erh-li-kang], Shangjie [Shang-chieh] graves, etc. But despite these twenty-nine years of study, we know little about this site, for hardly any information and no inventory have been published. As work stands at present, we may conclude that Zhengzhou was a large industrial and cultural centre under the Shang, with bronze-foundries and an industry in articles made of bone and horn. Pottery was also highly developed, its shapes anticipating those of bronze ritual vessels.

From a study of the strata of Yangshao until the Anyang phase, Zhang Guanzhi [Chang Kuang-chih] concluded that the cultural phase of Erligang dates to the period 1650 to 1400 B.C., while the Erlitou [Erh-li-t'ou] or archaic Shang phase dates from 1850 to 1650 B.C. However, at Erligang the casting of ritual vessels seems to have made its appearance late, probably at the end of the period.

Graves at Zhengzhou

Many burial sites have been excavated in this area: Erligang, Nanguanwai [Nan Kuan-wai], Baijiazhuang [Pai Chia-chuang], Renmin gongyuan, Minggonglu [Ming kung-lu] and Luodamiao [Lo Ta-miao]. There are hundreds of these graves which may be divided according to shape and use into three distinct groups:

Small graves: The commonest and most numerous graves at Zhengzhou are irregular in form. They may contain one or several human skeletons but no grave-goods.

Other small graves: these are rectangular in form, measuring 2 by 0.5 metres with a depth of 1.40 metres. They contain funerary urns made of grey pottery that hold ashes and calcined human bones. In the centre is a small pit called *yaokeng [yao-k'eng]* ('waist pit') contain-

ing a dog. This type of grave is believed to belong to the Middle Shang period.

Large tombs: these have been excavated mainly at Renmin gongyuan and Baijiazhuang; they are rectangular, measuring 3 by 1.20 metres by 2 metres in depth and possess a funerary chamber with *yaokeng* pit. The coffin was placed in the centre of the chamber with platforms overhanging it. Some of these graves belong to the Middle and others to the Late Shang phases.

Ritual Vessels and their Ornament

Nothing is yet known of the origins of this magnificent technique of bronze founding. However, large numbers of ritual vessels have been brought to light in recent years at Zhengzhou and at other sites in northern and central China. They are believed to be the earliest bronzes discovered so far, dating from the middle of the Shang dynasty. Thus they are the precursors of the Anyang vessels. The problem of dating these pieces is not straightforward however, and Sueji Umehara and Takeshi Sekino, for example, rule out the possibility of so early a date. In their opinion, these objects reflect a decline in the art of bronze, and they place them at the end of the Shang or at the beginning of the Zhou [Chou] dynasty.

Thanks to the work of recent years and the study of the method of casting, the problem can now be resolved. The opinion of these two eminent scholars is considered to be too categorical, and the earliest ritual vessels from Zhengzhou are dated to the end of the sixteenth or beginning of the fifteenth century B.C. In the light of present knowledge of the vessels of the Middle Shang period, and more particularly after the excavations of 1974 at Panglongchen [P'ang Lung-ch'en] in Hubei [Hupei] province, where over 160 bronzes were unearthed, most scholars agree that bronze casting was already in a very advanced state of development by this period.

Although they vary considerably, the shapes of bronze vessels seem to have been limited to the *ding, li* and *lihe, pan, jia, jue, fang ding, gu, xian, yu, lei* or *zun.* These vessels are characterized by thinner walls and casting inferior to that of the Anyang vessels. They are also less powerful and less sure in shape, but they are elegant, possess a real charm and are therefore magnificent objects.

None of the pieces scientifically excavated has an inscription, but this is insufficient proof that the Middle

7
Pan with a band of stylized *kui* dragons round the outside.
Bronze
Height 10.5 cm.; diameter 30 cm.
Excavated in 1955 in tomb No. 2 at Baijiazhuang, north-eastern quarter of Zhengzhou

Shang bronzes were not inscribed. Two vessels in private collections, which, on grounds of shape, ornament and material may be attributed to the Zhengzhou period, bear inscriptions of three characters. Since the pieces were not excavated scientifically, it is impossible to affirm or deny that the inscriptions date from the Zhengzhou period or were added later.

The highly evolved decoration of the vessels from Zhengzhou is almost always engraved. Geometric in outline, the decoration consists either of fine lines heightened with circular motifs and zigzags or of *taotie* *[t'ao t'ieh]* masks and spirals in relief. The *taotie*, sometimes without eyes, is repeated twice and often placed at the centre of the spirals. On either side of this frieze, there is usually an ornament consisting of small circles delimited by two raised lines.

8

Anyang

Anyang or Yinxu was the last Shang capital built by King Pan Geng (nineteenth king of the Shang dynasty), the period considered by historians to have been the transitional phase between the periods which they call respectively Shang and Yin. Indeed, reference to the Chinese classical and historical texts shows that, at the accession of King Pan Geng to the throne, the Shang kingdom was in a phase of decline. To reverse the trend and give new life to the dynasty, the king decided to reside at the site of the ancient capital of Tang 'the Victorious' (founder of the dynasty). Accordingly he crossed the river and built a city; this was Yinxu, which was to remain the capital of the kingdom to the end, i.e. for three centuries, from the end of the fourteenth or beginning of the thirteenth century to the eleventh century B.C. The relocation of the capital is the pretext on which historians re-named the Shang dynasty 'Yin'. There is no archaeological foundation or precise inscription on which to base this purely arbitrary change. The only writings of the period to have survived are the oracle inscriptions, and they do not mention the town of Yin or that name, nor does the character *yin* itself appear in these texts. Father J.-A. Lefeuvre suggests that the Zhou were the first to use the term *yin* to express disdain for the vanquished Shang. However, we shall continue to use the name Yinxu when we have to refer to this town or this site.

Excavations

Begun in 1928, the scientific excavations in these regions were centred mainly on the sites of Xiaotun [Hsiao-t'un], Hougang [Hou-kang], Sipanmo [Ssu P'an-mo], Houjizhuang [Ho Chi-chuang], Xibeigang [Hsi Pei-kang], Wu guancun [Wu Kuan-ts'un] and Dasikongcun [Ta-Ssu K'ung-ts'un]. The results have always been extraordinarily fruitful, and they multiply as new discoveries are made every year. The most characteristic remains are those of a cemetery containing ten royal tombs and hundreds of other burial sites. The most interesting discovery from the architectural point of view was a series of foundations of royal buildings — walls and a terrace — made of superimposed layers of rammed earth known as *hangtu*.

8
General view of Royal Tomb HPKM 1500 excavated at Anyang
Institute of History and Philology, Academia Sinica, Taipei

9
Zun with decoration in three registers: pairs of confronted *kui* dragons at the base of the neck and on the foot; *taotie* masks composed of two confronted dragons on the central swelling. The four triangular blades on the neck filled with cicada-wing motifs are composed of *taotie* masks.
Bronze, green patina with touches of cuprite
Shang dynasty, Anyang period, thirteenth-eleventh century B.C.
Height 32 cm.
Museum Rietberg, Zurich

9

10

11

10
Fang ding with a stag's head, *kui* dragons and birds of prey, against a background of *leiwen,* on all four sides.
Bronze
Height 60.5 cm.
From Royal Tomb HPKM 1004, Anyang
Institute of History and Philology, Academia Sinica, Taipei

11
Bu with a round belly on a circular foot.
Bronze with a smooth deep-coloured patina
Shang dynasty, early Anyang period, fourteenth-thirteenth century B.C.
Height 22.5 cm.; diameter 28.5 cm.
Musée Guimet, Paris

12

took place between 18 March and 24 June 1936, in the circular pit H 17, archaeologists collected 17,096 inscribed pieces that represent the earliest written documents to have come down to us so far.

The Graves

Of all the architectural remains the graves and *keng [k'eng]* pits are the most attractive from the point of view of their archaeological content. The pits, the sole purpose of which was to hold sacrificial creatures (birds, animals, prisoners, etc.), are usually square or rectangular. On the basis of their contents they may be divided into: pits for chariots, some of which contained the charioteer and horses and all their trappings; pits for horses, sometimes with a groom; pits for oxen, sheep, pigs, birds and dogs; pits for human victims.

The contents of these *keng* pits are as nothing compared with the extraordinary riches of the royal tombs discovered at Xibeigang, Wu guancun and the twenty post-Shang graves of Luoyang, Hougang.

Hougang: This first royal Shang grave to have been found at Yinxu is a rectangular pit measuring 7 by 6.2 metres by 8.5 metres in depth, with two access ramps. The funerary chamber at its centre measures 5.7 by 4.4 metres and is 1.5 metres in depth and includes a *yaokeng* pit. Although it had been despoiled before it was scientifically excavated, this small royal grave yielded nearly two thousand grave-goods made of gold, bronze, jade, bone, pottery, etc.

Wu guancun: This royal tomb was discovered in 1950 one kilometre north of Wu guancun; it is a rectangular pit oriented north-south and measuring 14 by 12 metres at its highest point and 7.20 metres in depth. Its length is extended by two access paths, each about 15 metres long. It had been despoiled at least twice before it was discovered, and the grave-goods were fairly meagre,

A variety of objects have been unearthed, including stone, sculptures in white marble, engraved jades, articles made of shell, bone and horn, fragments of articles made of bronze and pottery — grey, black and especially white. The decoration of the white pottery, which was not even known to exist prior to these campaigns, is particularly interesting. The art of bronze also makes an honourable showing with very large numbers of ritual vessels and weapons. Nevertheless, from the historical point of view, the greatest revelation of these excavation campaigns was that of the royal archives inscribed on bone and tortoise-shell. During the thirteenth excavation campaign, which

13

comprising in particular white pottery shards and a large stone drum with a boldly engraved figure of a stylized tiger. The walls of the funerary chamber were made of tree-trunks; the imprint is all that remains visible of the former wooden floor.

Xibeigang: This royal cemetery composed of ten large tombs was discovered in 1934 and 1935. Large pits (approximately 20 by 15 metres and 10 to 12 metres in depth) are hollowed out of the ground; they are either square, rectangular or cruciform in plan. At the bottom is a wooden funerary chamber of the same shape as the

13
Pan decorated inside with a coiled dragon whose head resembles a *taotie;* on the inner edge a series of silkworms are depicted.
Bronze with a green patina
Shang dynasty, thirteenth-eleventh century B.C.
Height 11 cm.; diameter 33.5 cm.
Simone and Alan Hartman Collection, New York

grave. These tombs, known to Chinese archaeologists as 'large tombs' or 'royal tombs', have two, three or four ways of entry; these may be steps or ramps which end

14

14
Ding, decoration in two zones consisting of stylized *kui* dragons on
the neck and a geometric motif with raised dots on the body.
Bronze with light-coloured patina
Shang dynasty, thirteenth-eleventh century B.C.
Height 19.5 cm.
Alan and Simone Hartman Collection, New York

15
Li, tripod vessel, with three large *taotie* masks.
Bronze with a green and red patina
Shang dynasty, thirteenth-eleventh century B.C.
Private collection

34

15

16
Jia, with *taotie* masks composed of confronted *kui* dragons on neck and body.
Bronze with green patina
Shang dynasty, Anyang period, thirteenth-eleventh century B.C.
Height 39.5 cm.
Museum für Ostasiatische Kunst, Cologne

17
Jue, flat-based, the body, neck and legs decorated with an all-over design of *taotie* masks and stylized cicada-wings.
Bronze with green patina
Shang dynasty, Anyang period, thirteenth-eleventh century B.C.

Height 22.5 cm.
Museum für Ostasiatische Kunst, Cologne

18
Jia with a band of confronted *kui* dragons on the belly, forming a *taotie* mask on a ground of *leiwen*. The caps and neck are decorated with blades, or stylized cicadas, containing spirals; the handle at the side has a *taotie* mask in high relief.
Bronze
Shang dynasty, thirteenth-eleventh century B.C.
Height 33 cm.; diameter 17.5 cm.
Private collection

16

17

18

either at the bottom or at a sort of terrace situated at the top of the funerary chamber. However, the approaches to tombs HPKM 1217 and HPKM 1500 present unusual features: the western ramp of tomb HPKM 1217 forms a right angle, while the northern approach of tomb HPKM 1500 has two lateral passages adjoining the mouth.

A large number of sculptures in white marble were brought out of these tombs (most of them from tomb HPKM 1001), including two fragments of a human bust (tomb HPKM 1004), decorated bones, bone vessels, white pottery, bronze weapons (mainly *ge [ko]* halberds) (tomb HPKM 1003), bronze spear-heads and warriors' helmets (tomb HPKM 1004), bronze ritual vessels, jades and articles made of shell, etc.

These ten tombs, with that of Wu guancun, may be those of the eleven Shang kings from Pang Geng to Di Yi (last but one of the Shang kings).

Xiaotun: A pit measuring 5.6 by 4 metres and 6.2 metres in depth was found in 1976 to the north-west of Xiaotun. The wooden funerary chamber of this tomb, to which the inventory number 'tomb 5' has been given, and which had not been despoiled before it was excavated, contained 1500 objects, of which 500 were made of jade and bone and 440 of bronze. It was possible to date this tomb from the inscriptions Fu Hao (Lady Hao) and *Simuxin [Ssu-Mu-hsin]* on the vessels. In fact, Fu Hao is well known from the oracle inscriptions: she was probably the only lady to be raised to the rank of queen when she became the wife of King Wu Ding (twenty-second Shang king). Here, for the first time, is a Shang tomb to which a precise date can be given. On the basis of the chronology of Dong Zuobin, who dates the reign of Wu Ding to 1339–1281 B.C., and with the knowledge that Fu Hao was his wife, it may be assumed that this tomb dates from the beginning of the thirteenth century B.C.

Luoyang: During the excavations of 1952, twenty tombs were unearthed in the region of Luoyang. Having a rectangular chamber with a *yaokeng* pit, *hangtu* construction, paintings on wood and numerous offerings of dogs and men, they are typically Shang. The grave-goods comprise bronzes, pottery, jades, and articles made of shell. There is, however, a strong Zhou influence: after their victory, the Zhou had forced many Shang to emigrate to Luoyang, where they were influenced by Zhou art to some extent.

These tombs must be considered to be post-Shang; that is to say, they were built by survivors of the Shang, but under the Zhou dynasty.

Ornament on Ritual Vessels

Evolution of Ornament: The Shang decorative style is not unique to the bronzes; the same ornament is found on jades, bones, sculptures in white marble and on white pottery. The art of bronze is merely an extension of these other arts.

In the course of the four centuries during which the Shang occupied Yinxu, ornament and the way in which it was used underwent certain changes. In the early period, the decoration covered the whole vessel, usually with extended forms of the *taotie*-mask motif. Later on, *kui [k'ueh]* dragons made their appearance at the neck and on the foot of the vessel. By the end of the Anyang phase, the *taotie* had become completely dissolved—on an unvarying ground of *leiwen* spirals.

Decorative Motifs: Whether a vessel has no decoration or a linear pattern of one more fillets or is covered with highly developed and complex ornament such as the *taotie*, balance and originality are the two principal characteristics that give Shang art its strength. The shapes of the motifs, which are either realistic or figurative, but for the most part highly stylized, are composed of almost unvarying geometric and linear elements and of zoomorphic motifs.

Geometric and Linear Motifs: These consist of lozenges, whorled circles, spirals (often angular), meanders resembling the Greek T-shaped motif and the *leiwen*, which is a combination of spirals and meanders. The latter occurs either as the principal element of the decoration or as a background, emphasizing the zoomorphic motifs and accentuating the relief.

Zoomorphic Motifs: These too occur in great variety, but the two principal motifs are: a) the *kui* dragon in profile, its eye, mouth, crest, tail, one paw with a claw and (sometimes) a spur visible. In very rare instances, it assumes the appearance of an elephant; when it does so it has a sort of trunk as an extension of the mouth; b) the

19
Xian with a band of birds above, on a ground of *leiwen*; on the feet, *taotie* masks in low relief.
Bronze with a green and red patina
Late Shang dynasty, eleventh century B.C.
Height 42 cm.
Tai Collection, New York

19

20
Shao, with decoration of geometric motifs, whorled circles and winged creatures on a background of *leiwen*.
Bronze with rough patina
Shang dynasty, Anyang period, thirteenth-eleventh century B.C.
Height 16.5 cm.
Musée Guimet, Paris

20

21

21
Guang with a winged creature or bird of prey, the wings of which cover the whole of the front part of the vessel, and dragons, on a background of *leiwen*. The handle represents a bird with an elaborately hooked beak and two prominent horns.
Shang dynasty, Anyang period, thirteenth-eleventh century B.C.
Height 22 cm.; length 32.2 cm.
Metropolitan Museum of Art (Rogers Fund 1943), New York

22

22
He with a handle and a spout; the cover has a ring and is attached to the handle by a short chain; this piece is ornamented with stylized dragons.
Bronze with a greyish-green patina encrusted with malachite and azurite
Shang dynasty, thirteenth-eleventh century B.C.
Height 26 cm.
Museum Rietberg, Zurich

23
Fang yi, taotie masks in low relief on a background of *leiwen* on all sides and on the cover.
Bronze with green patina in which there are traces of azurite and malachite
Shang dynasty, thirteenth-eleventh century B.C.
Height 22.5 cm.
Frau Dr Emma Gross Collection, Zurich

24

taotie mask, often composed of two confronted and symmetrical *kui* dragons in profile, each constituting half the mask. The eyes, eyebrows, horns, jaw and what may be called the ears are represented either as compact elements or dissolved, i.e., separated from each other.

In addition to these two basic motifs, which occur on most Shang bronzes, many other animal figures are found: owls, the rarest of the animal figures; cicadas of two types, one representing the insect in a very stylized form, the other in a triangular form resembling a knife blade; silkworms, fish, frogs, tortoises, hares, snakes with triangular heads and undulating bodies, often presented frontally; elephants, in profile (this figure is extremely rare); wild beasts, usually tigers; cervidae, especially rams but showing only the heads and coiled horns; bovidae, especially buffalos with trimmed horns; birds, the domestic sort and birds of prey and human figures.

Zoomorphic ornament, which often appears in relief against a background of *leiwen*, is sometimes treated in middle relief, thus enhancing the artistic quality of the object.

24

Fang jia or square *jia* with a cover surmounted by two birds in full relief; the two narrow ends are decorated with an owl; decoration on the other two sides in three zones consisting of *kui* dragons, a band of whorled circles and a large *taotie* mask; blade decoration on the legs.
Bronze with olive-green patina
Shang dynasty, Anyang period, thirteenth-eleventh century B.C.
Height 30.5 cm.
Albright-Knox Art Gallery, Buffalo, N.Y.

The Zhou Dynasty

Chronology of the Zhou [Chou] Dynasty

Western Zhou or Xi [Hsi] Zhou: 1111–770 B.C.
   Capital Haojing [Hao-ching]

Eastern Zhou or Dong [Tung] Zhou: 770–256 B.C.
   Capital Luoyang

Period of the Spring and Autumn Annals or Chunqiu [Ch'un-ch'iu]:   722–481 B.C.

Period of the Warring States or Zhanguo [Chan-kuo]:   453–221 B.C.

Zhou Kings (dates according to Dong Zuobin)

Wu Wang	1111 B.C.	Hui Wang	676 B.C.
Cheng Wang [Ch'eng Wang]	1104 B.C.	Xiang Wang [Hsiang Wang]	651 B.C.
Kang Wang [K'ang Wang]	1067 B.C.	Qing Wang [Ch'iang Wang]	618 B.C.
Zhao Wang [Chao Wang]	1041 B.C.	Kuang Wang [K'uang Wang]	612 B.C.
Mu Wang	1023 B.C.	Ding Wang [Ting Wang]	606 B.C.
Gong Wang [Kung Wang]	982 B.C.	Jian Wang [Chien Wang]	585 B.C.
Yi Wang (Jian [Chien])	966 B.C.	Ling Wang (Qiu [Ch'iu])	544 B.C.
Xiao Wang [Hsiao Wang]	954 B.C.	Jing Wang (Gai [Ching Wang (Kai)])	519 B.C.
Yi Wang (Xu [Hsü])	924 B.C.	Yuan Wang [Yüan Wang]	475 B.C.
Li Wang	878 B.C.	Zhending Wang [Chen-ting Wang]	468 B.C.
Cong He [Ts'ung Ho]	841 B.C.	Kao Wang [K'ao Wang]	440 B.C.
Xuan Wang [Hsüan Wang]	827 B.C.	Weilie Wang [Wei-lie Wang]	425 B.C.
You Wang [Yu Wang]	781 B.C.	An Wang	401 B.C.
Ping Wang [P'ing Wang]	770 B.C.	Xian Wang [Hsien Wang]	368 B.C.
Huan Wang	719 B.C.	Shenjing Wang [Shen-ching Wang]	320 B.C.
Zhuang Wang [Chuang Wang]	696 B.C.	Nan Wang	314–256 B.C.
Xi Wang [Hsi Wang]	681 B.C.		

Zhou royal house suppressed by King Zhaoxiang [Chao-hsiang] of the Qin [Ch'in] dynasty   256 B.C.

Foundation of the empire by King Zheng [Cheng] of the Qin dynasty, who proclaimed
himself Qin Shihuangdi [Ch'in Shih-huang-ti]
                                                                                                221 B.C.

GENERAL REMARKS

The Zhou dynasty is the longest in the history of China. It continued for over eight centuries from the eleventh to the third century B.C. Despite the longevity of the dynasty, it is proportionally better documented than the Shang. There are various sources, including a series of books which survived the holocaust of 213 B.C., a large collection of inscriptions on bronze and stone as well as Shang oracle inscriptions.

The *jiaguwen* provide a vast amount of information on the pre-dynastic period. They tell us that the Zhou, who were non-Shang by race but had been assimilated, bore the title *hou* (marquis) and paid tribute to the Shang kings. Shang Ladies *(Fu-X)* were given to Zhou leaders in marriage. The Zhou developed in parallel with the Shang, whom they gradually supplanted, thanks to their military strength. The cultural and military rivalries between the two races culminated during the reign of King Di Yi: the Zhou attacked and destroyed the Shang kingdom. The date of the overthrow of the Shang dynasty is extremely controversial: the traditional date is 1122, Dong Zuobin gives 1111, the *Bamboo Books* 1047, Karlgren 1027, Yetts 1050, etc.

When the new Zhou dynasty was founded, Zhou culture was comparable with, if not slightly superior, to that of the Shang. Being a tributary and possessing relatively important power and lands, the Zhou assimilated the art and civilization of their predecessors, subsequently developing and modifying them as a consequence of provincial influences. There is no longer any shadow of a doubt that the Zhou were familiar with the technique of bronze casting before their dynastic phase. The *gui* known as the Malcolm *gui* or Kang Hou *gui* (now in the British Museum), with its inscription dating from the earliest years of the Zhou dynasty and recounting the second Zhou attack on the Shang capital, is ample evidence that this was so. The inscription on it may be translated: 'The king attacked the Shang capital, then ordered Kanghoubi [K'ang Hou-pi] [to remain] at Wei. Mu Zitou [Mu Tzu T'ou], Ni and Bi [Pi] had this precious sacrificial vessel made for their father. [Signed: clan-name]'. Wei is the name of the state governed by Bi, Marquis of Kang, a region that embraced the lands of Anyang, the old royal Shang capital. The inscription, therefore, sanctions the assumption that the vessel was cast after the victory of King Chen Wang during the Shang rebellion, which, according to historical texts, took place in the second year of the reign of Cheng Wang, that is, in about 1102 B.C. (date calculated from Dong Zuobin's chronology). But without entering into a discussion of the precise date of the piece, one can say that it was cast in the earliest years of the Zhou dynasty

25
Fang lei with flanges, confronted *kui* dragons on the neck and foot, *taotie* masks and confronted *kui* dragons on the body.
Bronze, green patina with traces of cuprite
Beginning of the Zhou dynasty, eleventh-tenth century B.C.
Height 62.7 cm.
St Louis Art Museum, St. Louis

26

27

26
Zun, the belly decorated with birds with long retroflected tails; a band of confronted birds with blade motifs above on the neck.
Bronze with a sea-green patina
Beginning of the Zhou dynasty, eleventh-tenth century B.C.
Height 16.4 cm.
Art Museum, Princeton University, Princeton, N.J.

27
Fang ding, the four corners decorated with confronted birds in low relief against a background of *leiwen;* the four legs represent similar birds in a standing position.
Bronze
Beginning of the Zhou dynasty, dated *c.* 1024–05 B.C.
Height 24.5 cm.; width 22.5 cm.
Asian Art Museum, San Francisco, Avery Brundage Collection

28
Fang yi with flanges, sides slightly bulging; decoration of dissolved *taotie* masks and bands of *kui* dragons on a background of *leiwen* on both vessel and cover.
Bronze
Beginning of the Zhou dynasty, eleventh century B.C.
Height 28 cm.; length 17.5 cm.; width 15 cm.
28   Metropolitan Museum of Art (Rogers Fund 1943), New York

29

29
Group of twelve ritual vessels and bronze table, decorated with *taotie* masks, dragons and birds.
Provenance: Fang xiang, Shaanxi province (1901)
Bronze
Shang and Zhou dynasties, eleventh-tenth century B.C.
Metropolitan Museum of Art (Munsey Bequest 1924), New York

by the same technique as the Shang vessels. Such technical mastery could not have been assimilated in a few years, and it is extremely likely that the Zhou bronze-founders were making pieces of this sort well before the dynastic period. This supposition was confirmed after 1958 by the discovery of several bronze ritual vessels at Qishan [Ch'i-shan] in Shaanxi province. The most interesting piece excavated at this site, which is that of the proto-Zhou capital, is known as the Zhen *gui* [Cheng *kuei*]. This piece may be dated by the year of King Wu Wang's victory over the Shang. Zhen, who was presumably an important person in the kingdom, was privileged to be present at the commemorative ceremony held by the king in honour of the dead king, Wen Wang. Zhen had this bronze cast in memory of that event. The *gui,* which measures 24 centimetres in height and has a diameter of 20.5 centimetres, has four handles and a cubic base. This shape of *gui* was

entirely unknown under the Shang. The principal ornament is also peculiar to the Zhou and consists of an animal with the head of a bird that has no known counterpart in the iconographic repertoire of the Shang. The inscription, which is in a literary style comparable to that of the *Shijing [Shih-ching],* consists of seventy-eight characters narrating the Zhou victory. All the vessels collected at Qishan were cast in the same way as they would have been under the Shang, i.e. in piece-moulds.

Like the bronze vessels of the Shang era, the Zhou vessels were inspired by or copied from pottery originals. However, shapes became modified under the new dynasty. This may have been due to a new fashion, an adaptation to a new use, or it may have reflected local taste. Yet the arrival of the Zhou was attended by no out and out break in the tradition of bronze making: the changes in styles and shapes occurred gradually. This slow iconographic mutation partly explains the difficulty of dating pieces described as 'traditional', i.e. those made from the end of the Shang to the beginning of the Zhou dynasty. Moreover, the inscriptions show that the nomenclature of Zhou ritual vessels did not follow hard and fast rules; thus the same vessel could have different names at different periods, or vessels of the same name might assume various shapes in obedience to local convention or to their diverse uses.

31

30
Zun with birds on a ground of *leiwen* in the middle of the body.
Bronze with a green and red patina
Late Shang dynasty or early Zhou dynasty, eleventh-tenth century B.C.
Height 25.2. cm.
Simone and Alan Hartman Collection, New York

31
Hui or *gui.* The cover has two lateral ears projecting from the jaws of animal masks; the edge of the cover, the neck and the foot are ornamented with a band of *taotie* masks.
Bronze with a sea-green patina
Late Shang dynasty or early Zhou, eleventh-tenth century B.C.
Height 22 cm.; diameter 28 cm.
Private collection

WESTERN ZHOU

The beginning of this period, corresponding to about the eleventh century B.C., is extremely confused as far as the art of bronze is concerned. Not only did Shang bronzes survive, for a considerable number of bronze-founders settled at Luoyang (which explains why there is no yardstick by which to distinguish late Shang from post-Shang pieces), but also the bronzes of the early Zhou were influenced by or were straightforward continuations of Shang vessels.

Morphological and iconographical changes came slowly and gradually. Thus from the end of the eleventh or beginning of the tenth century B.C. a number of Shang shapes and characteristics were abandoned. This was the case with the vessels *gu* and *jue* and with the bell *nao*, which was no longer made, as well as with the *zun* and *yu*, which were employed less commonly. In other cases, considerable morphological changes resulted in the creation of new and typically Zhou shapes:

the *gui*, which had had a ring foot under the Shang, was now placed on a large square stand and given a lid;
the *guang* lost its lid and became a pouring vessel termed *yi*;
the *pan* acquired handles and, in about the eighth century B.C., the ring foot was replaced by three small feet;
the dipper *shao* was also modified, the stem becoming longer and curved;
new forms, such as the *fu*, made their appearance.

From the point of view of iconography, most of the decorative elements from the beginning of the Western Zhou were to all intents and purposes similar to those of the Shang, or the differences, where they occurred, were negligible. The design was symmetrical at the centre and most commonly represented the *taotie* mask.

From the tenth century B.C. onwards, or perhaps towards the end of the reign of Cheng Wang, a new and highly individual style made its appearance. The evolution of ornament as it freed itself from its iconographic content was expressed in a reduction in the number of zoomorphic motifs. Animal ornament and the *taotie* mask disappeared fairly rapidly. The backgrounds of engraved *leiwen* were abandoned, and the principal motifs stood out against an undecorated surface. A new iconographic repertoire made its appearance; it was more varied and was composed of more complex motifs, including horizontal fluting, scales, wave patterns, interlacing, spirals, hooks, triangles and curves, etc. This repertoire was to remain in widespread use from the beginning of the Eastern Zhou.

The inscriptions also changed. The longer texts, written in the *dazhuan [ta-chuan]* or 'great seal' style, refer to military events, rewards, ceremonies, nominations of feudal lords or officers.

In his book *Chou China* Zheng Dekun [Cheng Te-K'un], like Muzuno Seiichi in *Bronzes and Jades of Ancient China*, divides Western Zhou ritual vessels into three groups:

a) early Western Zhou: this period is characterized by the pieces unearthed at the sites surrounding the old royal capital in Shaanxi province and at Luoyang. The vessels

32

32
Yu with ear handles; the ring foot and the neck are decorated with *kui* dragons confronted to form a *taotie* mask in high relief; the blade motifs on the belly enclose stylized *taotie* masks.
Bronze with green patina
Beginning of the Zhou dynasty, eleventh-tenth century B.C.
Height 41.8 cm.; width 56.5 cm.
Freer Gallery of Art, Washington, D.C.

33
You with a band of confronted birds on the foot, cover and body, on a ground of *leiwen*; the handle is adorned with *kui* dragons on a ground of *leiwen* and terminates in two animal masks in full relief.
Bronze with a black and green patina
Late Shang dynasty or early Zhou, eleventh-tenth century B.C.
Simone and Alan Hartman Collection, New York

33

34

34
Pan, the ring foot with *taotie* mask supports the cup, on the outside of which is a band of *kui* dragons interrupted by *taotie* masks in high relief; the rim of the cup is surmounted by six birds in full relief; the inside is decorated with dragons, fish and birds.
Bronze with green patina
Beginning of the Zhou dynasty, tenth-eighth century B.C.
Height 15 cm.; diameter 35 cm.
Private Collection, Paris

35

35
Lei with high shoulders decorated with five bands of interlacing dragons.
Bronze with green patina
Middle Zhou period, eighth century B.C.
Height 30 cm.
Alan and Simone Hartman Collection, New York

belonging to a batch often occur in pairs. Both Shang and Zhou ornament is present at this period. The commonest motifs are the *taotie* mask, the *kui* dragon, stylized cicadas and birds. The inscriptions are of some length and often relate to the reigns of Cheng Wang and Kang Wang.

b) Middle Western Zhou: illustrated by finds in the environs of Xi'an [Hsi'an]. As a whole these pieces differ from the previous style; only a few rare Shang decorative motifs remain. Purely Zhou shapes and motifs are by far the most numerous. Scales, fluting, etc. multiplied, while dragons, birds and the *taotie* masks became more stylized. These pieces may be dated from the inscriptions to around the time of the reign of Mu Wang.

c) Late Western Zhou: the last vestiges of Shang influence had disappeared, and the shapes and ornament of the ritual vessels of this period are typical of Zhou art. All these elements, except for a few variants, will be found under the Eastern Zhou and will influence the Period of the Spring and Autumn Annals.

In brief, there are five major elements that characterize the era of the Western Zhou: most of the wine vessels disappeared; the morphology of the ritual vessels changed; the principal decorative motifs appeared against a plain background; the commonest motif of the transitional phase was a bird with a long tail and an elaborate crest; new, mostly geometric, decorative motifs made their appearance.

EASTERN ZHOU

Under the Eastern Zhou, the evolution of the art of bronze was accompanied by new developments in the technique of casting. There were four different techniques by which legs, ears and handles could be attached to the body of the vessel: they could be cast in one with the body of the vessel, as in the earlier periods; they could be cast separately and then assembled in a mould of the body of the vessel; the body could be cast on its own, and the moulds for the accessories subsequently assembled round the vessel in their final positions; (The molten metal poured into the moulds ensured that feet and handles adhered firmly.) the body and accessories could be cast separately and then soldered to form the finished vessel.

36

36
Niaozun or ritual vessel of the *zun* type in the form of a bird.
Bronze
Zhou dynasty, eleventh-eighth century B.C.
Height 22.5 cm.
Yale University Art Gallery (gift of Mrs William H. Moore for Hobart Moore and Edward Small Moore Memorial Collection), New Haven, Conn.

The Eastern Zhou period proper lasted for a mere fifty years or so, too short a time to create a style of its own. The art of bronze was similar to that of the end of the Western Zhou. The only features of the impoverished and all but abstract ornament were undulating bands of ribbons, bands decorated with scales or waves and geometric motifs that stood out against undecorated surfaces. Animal ornament had practically disappeared but occurred in the form of heads sculpted in the round at the points where feet and handles were attached to the body.

PERIOD OF THE SPRING AND AUTUMN ANNALS
CHUNQIU [CH'UN-CH'IU]

Characterized by a proliferation of estates anticipating the constitution of a feudal society, this period is marked by a great variety of forms and regional styles. The art of bronze, which simply continued the tradition of the Western Zhou, was open to influences from north-west China. These were reflected in a distinct change of style. Ornament became more severe and was increasingly dominated by patterns of curves, mainly of the *qiequ [ch'ieh-ch'ü]* and *yunwen [yün-wen]* types. Furthermore, in the Period of the Spring and Autumn Annals, during which the use of bronze coinage became general, many merchants grew rich. Thus a new moneyed and land-owning class arose. Emulating the senior officials and the high aristocracy, its members had themselves buried with bronze ritual vessels and various other objects of every-day life. The fact that this privilege, which had been reserved under the Shang and the Western Zhou for the royal family and the high dignitaries of the state, had now become general, may perhaps explain the evolution and, especially, the decline in style and casting.

Scientific excavations from 1955 onwards have brought to light many tombs of the Period of the Spring and Autumn Annals containing vast numbers of bronze vessels. Seventy-four tombs were unearthed at Luoyang and 243 at Shangcunling [Shang Ts'un-ling], in Henan province. But it is the tomb of the Marquis of Cai [Ts'ai] and the finds at Liyu [Li-yü] that provide a conspectus of the shapes and art of the period.

The tomb of the Marquis of Cai *(Caihou) [Ts'ai-hou]*, studied in 1955, was constructed between 493 and 447 B.C. The 123 bronze vessels found therein revealed

37

37
Zhong, a bell with a circular handle decorated with interlacing stylized dragons; nipples in groups of three on the body.
Bronze
Zhou dynasty, tenth-eighth century B.C.
C.T. Loo Collection, Paris

38
You. An animal mask in full relief adorns the end of each of the two handles.
Bronze with a sea-green patina
Zhou dynasty, tenth-eighth century B.C.
Private collection

40

Li with cover, decorated with an all-over design of interlacing bands suggesting dragons.
Bronze
Zhou dynasty-end of the Period of the Spring and Autumn Annals, sixth century B.C.
Height 23 cm.
Asian Art Museum, San Francisco, Avery Brundage Collection

are very close to those of the Marquis of Cai's vessels, places the Liyu bronzes with those from the end of the Period of the Spring and Autumn Annals.

Liyu Bronzes

Excavated in 1923 after a cliff-fall at Liyu in Shanxi [Shansi] province, about 60 kilometres from Datong [Ta-t'ung], more than twenty bronzes were acquired by the great Parisian dealer Wannieck. They were exhibited at the Musée Cernuschi in Paris in 1924. Some of them, in particular a fragment of a *pan*, a *fang ding*, a *yi* ewer, etc., are now preserved in the Musée Guimet, Paris. Most scholars at the time believed them to date from the Han or perhaps Qin [Ch'in] dynasties, that is, about the third or second century B.C. Following the discovery of the tomb of the Marquis of Cai and a comparative study of morphological and stylistic features, general opinion nowadays is that the Liyu bronzes represent the art of the end of the Period of the Spring and Autumn Annals (722−481 B.C.).

From the point of view of style and casting technique, the bronzes form a coherent group. The pieces are light and very thin. The ornament is composed of several zones of different motifs and for the most part uses interlacing dragons or dragons twisted into spirals to form a garland round the vessel, triangles, whorled circles and interlace. Additional decoration is provided by zoomorphic ornament based on animal masks or in the form of buffalos, ducks, dragons, etc. in high relief and middle relief.

39

39

Dou with four side handles representing tigers in full relief; the whole vessel except the foot decorated with a design of interlacing dragons.
Bronze with light-coloured patina
Zhou dynasty − Period of the Spring and Autumn Annals, eighth-sixth century B.C.
Metropolitan Museum of Art (Rogers Fund 1925), New York

New Shapes

The first point to emerge in regard to the bronze ritual vessels of the Period of the Spring and Autumn Annals is that an immense variety of new and unusual shapes made their appearance. The following were unearthed: a *ding* with a spout; a *yiding* or *guangding*, a hybrid form, very

new and unusual shapes and the introduction of copper inlay. This new technique heralded the vogue for gold and silver inlays which were to become popular under the Warring States. The discovery of this tomb gave scholars the opportunity to revise the dates of the Liyu vessels, which in recent years had been thought to belong to the Warring States. A comparison of styles and shapes, which

40

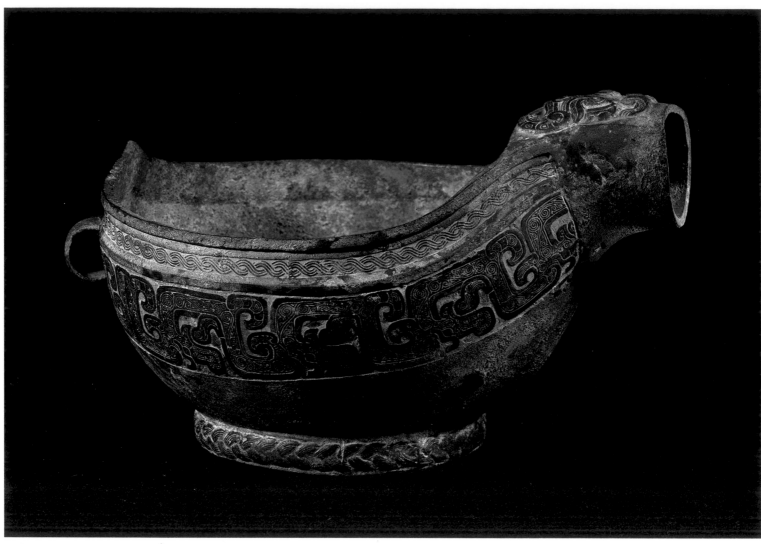

41

rare, combining two familiar types, having the legs of a *ding* and the silhouette of a *guang*; some large *fang hu*. They have a ring foot supported by two or four animals sculpted in the round, two handles at the sides and a cover with openwork ornament. The best-known example of this type is the *fang hu* brought out of the tomb of the Marquis of Cai. The other, more classical, shapes had also undergone slight morphological changes.

Iconographic Features

The iconographic repertoire of the Period of th Spring and Autumn Annals was strongly influenced by the style of the late Western Zhou and may be divided into three categories:

41
Yi ewer ornamented with interlaced dragons.
Bronze with a red patina
From Liyu
Late Period of the Spring and Autumn Annals or early Warring States, fifth century B.C.
Musée Guimet, Paris

42
Hu with braided ornament on the body; geometric motifs and two *taotie* masks in low relief on the neck.
Bronze with a green and red patina
Period of the Warring States, fifth-third century B.C.
Maurice Bérard Collection, Paris

a) zoomorphic motifs: *taotie* masks, *kui* dragons, *pan-long (p'an-lung)* dragons, *qiequ* curves; animal decoration (elephant heads, tigers, deer, fish, birds, serpentine interlacing dragons, etc.);

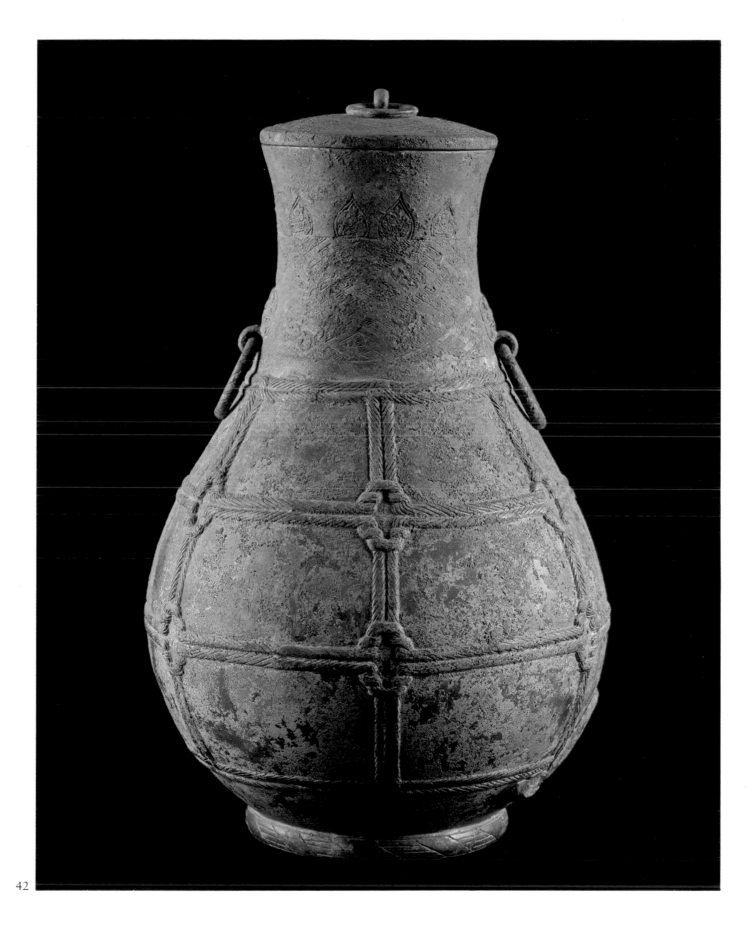

b) geometric motifs: *leiwen*, clouds and *yunwen* volutes, circles, spirals, rings, scales, flutes, triangles, etc.;

c) floral ornament: this occurs mainly on the covers of vessels and is composed of large openwork petals.

The commonest of all these ornamental motifs are the *qiequ* curves, *yunwen* volutes and *leiwen* spirals. It is also interesting to note that from the Period of the Spring and Autumn Annals onwards a new technique was employed for inscribing the vessels. This method, known as movable type, is based on the use of a block of movable characters that could be used in the casting of more than one vessel.

PERIOD OF THE WARRING STATES

During the Period of the Warring States, one of the most lively periods in the history of Chinese art, an admirable revival in the art of bronze took place. It manifested itself in the reappearance of the *taotie*, dragons, spirals, triangular blades, all interpreted with greater freedom, flexibility and richness. Once again the decoration tended to cover the whole surface of the vessel. Nevertheless, as time went on during the Period of the Warring States, a gradual decline in the art of bronze became apparent. This was due to the following phenomena: the introduction of bronzes meant solely for funerary use; a poorer alloy partly due to an increase in the lead content, a tendency to replace the bronzes by articles of the same shape made of pottery, over-refinement of decoration, the introduction and popularization of inlay, and the use of bronzes by a host of petty princes and lordlings.

The Shapes

Shapes inspired by and copied from pieces of the Period of the Spring and Autumn Annals evolved. Animal ornament was introduced as a consequence of contacts with small Sino-Siberian tribes and their influence. This had morphological repercussions in the human and animal figures that replaced the cylindrical or ring feet, the animals sculpted in the round that were used for handles and the use of chains instead of handles.

The ritual vessels of this period may be divided into two groups: gessels with highly evolved shapes and ornament and vessels lacking all ornament, or simply decorated, of coarse shape and indifferently cast.

43

43
Hu, with six bands, all different, representing interlacing dragons, birds and hunting scenes in which a man attacks a dragon, a tiger or a buffalo.
Bronze with dark patina
Zhou dynasty—beginning of the Period of the Warring States, fifth century B.C.
Height 25.5 cm.
Eskenazi Ltd., London

44
Detail of Plate 43.

45

45
Rectangular *ding*. The four feet are decorated with two bands of interlaced stylized dragons. There are two animals in middle relief on the cover.
Bronze with a green and red patina
From Liyu
Late Period of the Spring and Autumn Annals or early Warring States, fifth century B.C.
Height 16 cm.; width 25 cm.
Musée Guimet, Paris

46
Hu with five alternating bands depicting dragons or birds, inlaid in copper enriched with turquoise.
Bronze with a sea-green patina and touches of cuprite and azurite
Period of the Warring States, fifth-third century B.C.
Height 29 cm.
Dr. Franco Vanotti Collection, Lugano, Switzerland

Inlay

Inlay made its appearance at the end of the Period of the Spring and Autumn Annals, the technique becoming widespread under the Warring States. The taste for sumptuousness and rich decoration manifested itself in the use of wire made of gold, silver, copper, etc. to enhance and enrich the decoration of bronzes.

Motifs cast in relief were occasionally decorated with a carbonated substance with a quartz and cuprite base that scholars believe to have been lacquer. Much more rarely, decoration was painted or inlaid with glass and turquoise.

46

Iconographic Features

Despite the wide area covered by Chinese territory during the Period of the Warring States, there was an extraordinary homogeneity in the concept of ornament, and rich inlay was universally admired. The many ornamental motifs that had characterized the Period of the Spring and Autumn Annals continued to be employed under the Warring States, though in a more refined and a freer way:

a) Zoomorphic motifs: *taotie* masks, dragons, stylized cicadas, interlacing birds, serpentine dragons interlacing to form spirals, *panzhi (p'an-ch h)* draoons.

b) Geometric motifs: spirals, interlaces, volutes, whirling and interlacing curves. *Taotie* masks, sculpted in middle relief or more often in high relief, form the attachments for the rings and chains that replaced ears and handles under the Warring States. On the bodies and lids of vessels decorative bands were repeated and covered the whole surface. Animals sculpted in the round, or in middle relief sometimes decorated the lids of vessels or replaced handles.

c) Two innovations enlarged the iconographic repertoire of the Period of the Warring States: the first, which is simply an improvement of a technique employed at the end of the Period of the Spring and Autumn Annals, consists of damascened ornament in gold, silver or copper, often enriched with inlays of turquoise, malachite or other precious substances; the second, which is merely an application of the technique of inlaying, consists of representations of animals, scenes of everyday life and hunting scenes.

In the latter form of decoration the relief has disappeared altogether. The scenes occur in several registers and are often repeated, though with small variations.

Despite the richness and sumptuousness of ornamentation during the Period of the Warring States, this period marks the end of the art of bronze in China. The decline was due to the passing of the religious beliefs of the Shang and Early Zhou, to the appearance of Confucianism and Taoism and the attendant changes in Chinese society, and to the fact that these vessels came to be used solely for funerary purposes, whereas their original role had been purely ritual.

The art of bronze was to be more worldly under the Han, despite a lingering suggestion of features from the Warring States and the Zhou. Later, during the third to ninth centuries A.D., there would be a revival of the art, though in the totally different form of figures of divinities, Buddhist shrines, etc. in gilded bronze. But despite the quality and beauty of certain Wei bronzes, the technique never again approached the mastery of the Shang and the early Zhou founders.

Inscriptions

WRITING ON BONE AND TORTOISE-SHELL: *JIAGUWEN*

Study of the *jiaguwen* grew during the early years of this century prompted by Liu E and by the discoveries at Anyang, and thanks to important work by Dong Zuobin. These inscribed texts or divinatory statements concern sacrifices to the ancestors and to the spirits, military campaigns against the 'barbarian' tribes who surrounded the Shang kingdom, hunting, journeys, drought, rain, sickness, dreams, prophesies and other spiritual, natural and human affairs relating to the royal house or the kingdom. They represent a part of the royal archives produced some 3500 years ago and surviving in a fragmentary state to the present day. They are therefore highly important texts and are indispensable to the serious student of Shang society and, more especially, of the religion, institutions and social and political structure of the Shang world. Although they are still ignored by many sinologists, it is as well to remember that they are the only written documents of that ancient period to have been discovered so far.

The form of this archaic writing, consisting of between 2500 and 3000 characters, half of which stand for names of persons and places, was fairly unstable. Indeed, it has been established that the characters evolved and changed progressively during the last 273 years of the Shang dynasty. Furthermore, the interpretation of some 1500 characters is nowadays regarded as secure, while the meaning of the others remains extremely controversial. Yet despite this great gap in our knowledge, many fragments of texts have been deciphered and translated and have thus provided much information concerning archaic Chinese society.

Epigraphical study has shown that the characters may be divided into three groups. The largest consists of pictograms; these are representations of concrete objects, including animals and men and natural and man-made articles, such as weapons. They are crude and schematized images. The next includes ideograms or visual representations of scenes of daily life. These transpose symbols into ideas, suggesting an action or indicating a position. Finally, there are phonograms, the smallest category; these combine a pictogram with a phonetic element, utilizing a word with a similar sound to express a complementary idea. The fact that they developed late, mainly during the reigns of the two last Shang kings, explains their rarity in the *jiaguwen*.

Dong Zuobin has made an extensive study of the periodization, dating and chronology of the oracular inscriptions and has established a theory by which the material may be divided into five periods based on the following ten points:

1 Genealogy of the dynasty. The oracular inscriptions have enabled scholars to correct an error in the *Shiji [Shi Chi]* (Historical Records of Sima Qian [Ssu-ma Ch'ien]) by changing the order of the pre-dynastic ancestors and naming one more king.

2 Posthumous names. These were the ceremonial names given by the reigning king to a dead ancestor.

3 Name of the diviner. Most of the oracular sentences are of the following type: 'on such and such a day' [name supplied by two cyclical characters] divination [by such and such a diviner] questions [the oracle on this]:... [stating the question]'.

Dong Zuobin has observed that certain inscriptions mention several diviners. He has deduced that the diviners whose names were inscribed on one fragment were in all probability of the same period. By generalizing from this hypothesis Dong Zuobin has identified five distinct groups of diviners; these he has classified by reference to the posthumous names mentioned in the texts. These persons conducted their divinations in the name of the king, after they had inscribed the question. Study also reveals that they acted as annalists.

4 Find-place. After use, the inscribed pieces were sorted and arranged in pits specially intended for the purpose. Thus during the excavations of 1936 over 17,000 were discovered in pit H 17 at Yinxu. Dong Zuobin has observed that the find-places also corresponded to five periods.

5 'Barbarian' tribes. Chinese historical texts mention important military expeditions against the tribes that surrounded the Shang kingdom. These observations are confirmed, and detailed texts appear in the *jiaguwen*. Thus we learn that the armies of Wu Ding [Wu Ting] encountered the Tufang [T'u-fang], Erbo [Erh-po], Yifang, Bafang [Pa-fang] and Chifang [Ch'ih-fang]. Di Xin [Ti Hsin] conquered the Renfang [Jen-fang] and the Mengfang.

6 Names of officials. These were royal servants, ministers and other personages of the realm, the most important of

whom were the generals, princes and marquises. Study of these provides much information about the political structures of Shang society.

7 Types of divination. This ill-defined category brings in other criteria, but it is mainly those texts that relate to hunting and sacrifices which prove to be most instructive in the matter of periodization.

There are many references to hunting in the texts dating from the reigns of Wu Ding and Di Xin, and hunting was one of the reasons for levying troops. The inscriptions are of the type, 'on such and such a day, divination by such and such a diviner, question: if we go hunting at [place named] shall we catch anything? On that day's hunt, we caught: tigers [number], rhinoceros [number], deer....' On a single expedition the kill might amount to one tiger, forty stags, 164 foxes and 159 deer.

8 Words used. Of particular interest here are the predictions of fortunes in the coming decade. These texts are of the type, 'on such and such a day divination by [the diviner], question: will the coming decade be without ills?' The structure of the sentence varies depending on the period; sometimes the name of the month is also given, 'in this tenth month'. Sometimes a postscript announcing an event or a sacrifice or making a verifying statement is recorded on the bone. In the fifth period, new words make an appearance, and variants of the characters replace the earlier ones.

9 Style of the characters. This changed gradually from pictograms to phonograms in a development consequent upon increasing complexity of meaning. The transformation may be seen most clearly in connection with the cyclical characters, for which there are precise forms in each period. The development occurs and continues in the inscriptions on bronzes.

10 Calligraphic style. This differs from period to period. The predominantly large, bold characters of the first period had become small and painstaking by the fifth.

Study of the epigraphy of the *jiaguwen* is proving more and more essential to better deciphering of the inscriptions on bronzes, which are usually extremely brief, a fact that often leads to problems of interpretation. The oracular inscriptions are of some length; therefore, it is possible to decide whether a given character represents the name of a place or a person. Since the *jiaguwen* have been studied, it has sometimes been possible to interpret

the inscriptions on bronzes, for many names of individuals and places mentioned on the bronzes are also mentioned in the oracular texts.

WRITING ON BRONZE: *JINWEN [CHIN-WEN]*

This type of writing, which is much better known than the *jiaguwen*, is less interesting in so far as the Shang civilization is concerned but provides some information on the Zhou dynasty.

Vessels bearing inscriptions are fairly numerous: 55 per cent of *ding*, 65 per cent of *gui*, 52 per cent of *hu* and 59 per cent of *zun* are thought to be inscribed. Although over half the ritual vessels have *jinwen*, they are of practically no use in dating the bronzes, since it is often impossible to say whether a given vessel received its inscription at the time of casting or later.

Yet, as a result of the scientific excavations carried out during the past twenty-five years, we can now note certain differences between Shang inscriptions and those of the Zhou.

Inscriptions on Bronzes of the Shang Period

In the light of present knowledge of Chinese archaeology, scholars all agree that inscriptions on bronze did not make their appearance until the final phase of the Shang dynasty, in about the thirteenth century B.C. No inscriptions have so far been found on bronzes unearthed scientifically at Zhengzhou and dating to the end of the fifteenth or beginning of the fourteenth century B.C.

The inscriptions that were common under the Shang are in most cases brief formulae or sometimes consist of a

47
Zun with sloping shoulders and a wide mouth above a large belly; on the belly are three *taotie* masks on a ground of *leiwen*.
End of the Shang dynasty, twelfth-eleventh century B.C.
Height 52 cm.
Eskenazi Ltd., London

48

49

48
Jia, with flat base, band of stylized *taotie* masks on the body.
Bronze with dark green patina
Shang dynasty, Zhengzhou period, end of the sixteenth-fourteenth century B.C.
Height 22 cm.; width 16.5 cm.
Musée Guimet, Paris

49
Bu with three ram heads in high relief on the shoulder; the entire surface of the vessel is decorated in three zones: two with *kui* dragons and the third with *taotie* masks.
Bronze with a green patina
End of the Shang dynasty, eleventh century B.C.
Height 26.5 cm.; diameter 26.5 cm.
Musée Cernuschi, Paris

single ideogram. The characters are often similar in type to those of the oracular inscriptions. The characters are either inside the vessel, under the base, or under the handle in the case of the *jue* and the *jia*.

The formulae can contain: the name of the owner or of the clan in one or two characters; the verb *zuo* (made for); an indication of the ancestor for whom the vessel has been cast, giving his ritual title: *fu* (father), *zu [tsu]* (grandfather), *mu* (mother), etc., followed by the cyclical symbol: *jia [chia]*, *yi [i]*, *bing [ping]*, *ding [ting]*, etc.; the designation of the vessel, either by the expression: *zunyi*

[tsun yi] (sacrificial vessel), often preceded by the word *bao [pao]* (precious), or by the name of the vessel: *ding*, *jue*, etc; and, to complete the phrase, a signature representing a name or a clan.

However, the commonest Shang inscriptions are: 'to my father so and so', 'to my grandfather so and so', 'to my mother so and so,' 'by so and so to his father so and so,' and 'made by so and so for his father so and so'. There are also rare dedications to a mother, to a wife and to an elder brother, the last being by far the least common.

50

Marks

Another class of inscription on Shang vessels are those known as marks. They sometimes appear alone on a vessel, with no other inscription; sometimes they occur at the end of an inscription, more rarely at the beginning of an inscribed text. They are either simple or composed of several monograms.

According to Hayashi Minao marks may be divided into seven or eight hundred different types. The commonest is known as *xizusun [hsi tsu sun]*. Its meaning is very obscure, and the various interpretations are currently the subject of lively controversy. But Vandermeersch's hypotheses would seem to be the most plausible. He believes that the mark was used by the *zi* (princes), *xiaozi [hsiao tzu]* (squires) and *xiaochen [hsiao ch'en]* (pages) and was probably the emblem of a 'corps' of titled personages of various ranks. This category was probably identical with the *duozizu [to tzu tsu]* (corps of

princes) of the oracular inscriptions. The mark therefore represented an essentially military 'regimental corps'.

The second category of marks consists of the character *ya* and is known as *yaxing [ya-hsing]*. They resemble the cruciform plan of the Shang tombs and may be interpreted either as a symbol of death (?) or as a clan name (?). They often contain other characters that are place-names mentioned in the oracular inscriptions and believed to be seignorial houses. These marks resemble the Egyptian cartouches in method and content.

Hayashi Minao believes the *yaxing* to be family names rather than names of individuals, since they are used in different senses on vessels apparently dedicated to the same ancestor, who is called *fu* (father) on one and *zu* (grandfather) on the others. It may thus be a signature used by several generations. *Yaxing* are found in large numbers on vessels of the early Western Zhou; in this instance they must come from members of the Shang race who had rallied to the new kings. But on these post-Shang vessels made by the Shang under the Zhou dynasty, the phrases are of the following type: 'mark + has made this sacrificial vessel; may his sons and grandsons use it to eternity as a precious vessel'.

By the end of the Shang dynasty the inscriptions had lost their exclusively funerary meaning. They had become longer and recorded non-religious events in the presence or absence of the king. The phrases were of the following types: 'Ge [Ko] gave Feng a jade. Then [Feng] made this vessel for his grandfather Gui [Kuei]. [Mark]', 'Day *jihai [chi hai]*. The king gave cowries to X. It was at [name of place]. Then X made this sacrificial vessel for his father Ji [Chi]. [Mark] Ya Zhong [Ya-tsung]'.

51

Inscriptions on Bronzes of the Zhou Period

Under the early Zhou inscriptions became longer than they had been under the Shang, because the practice of writing was becoming more widespread. The inscriptions on vessels are written in typically Zhou prose, recording exploits of war, ceremonies, etc. The calligraphic style known as *dazhuan [ta-chuan]* or 'great seal' style is distinguished by small characters when they are composed of few strokes and large characters when they have to be built up from many strokes. From this period onwards the texts recount the circumstances in which a given vase had been cast:

1 Sacrifice to an ancestor. 'Precious ritual vessel made for Reqi. May his sons and grandsons use it perpetually for ten thousand years'. 'Precious ritual vessel made for Father Jia. May his sons and grandsons care for it for ever'. 'Precious sacrificial vessel made for Father Ding. May his sons and grandsons honour it and use it for a myriad years'.

2 Homage on the occasion of a religious ceremony.

3 Journeys. 'X ordered this vessel for use on journeys to be made for Father Ding. May his sons and grandsons use it perpetually for a myriad years'.

4 Military campaigns. '…Zhan [Chan]. In time past the dead king ordered you to aid Prince Y. I [the king] adhere to the order of the dead king, and I charge you to go to the aid of Prince Y and to see that Z's troops are destroyed'. 'In the 5th year, in the 3rd month, at the last quarter of the moon, on the day *gengyin [keng-yin]*, the king having initiated a repressive expedition against the Xianyun [Hsien yün], at Tayu [T'a yü], Xijia [Hsi Chia] followed the king, cut off heads and took prisoners alive, spared no effort and was not discouraged'.

5 Hunting expeditions. 'The king… at Yu, went hunting at X, [caught] 2 tigers, [then] made this precious sacrificial vessel for Father Ding'.

51
You: majestic birds with tall crests and long tails on cover and body; the points at which the handle is attached consist of animal heads in full relief.
Bronze with light-coloured patina
Beginning of the Zhou dynasty, eleventh–tenth century B.C.
Freer Gallery of Art, Washington, D.C.

52
You. The neck and cover have a band of birds on a ground of *leiwen*; the handle terminates in two *taotie* masks in middle relief.
Bronze with a patina resembling green jade
Early Zhou dynasty, eleventh–tenth century B.C.
Height 37.5 cm.
Private Collection, Paris

52

53

53
Guang with cover in the form of a horned beast; the body of the vessel is decorated with a large bird against a plain background.
Bronze
Beginning of the Zhou dynasty, eleventh-tenth century B.C.
Height 24 cm.; width 29.5 cm.
Asian Art Museum, San Francisco, Avery Brundage Collection

54
Fang ding. The ribs of each side are decorated with a stylized bird on a ground of *leiwen*.
Bronze with a green patina
Early Zhou dynasty, tenth century B.C.
Height 17.5 cm.; length 15.5 cm.; width 12 cm.
Tai Collection, New York

55

56

55
Fang hu with cover surmounted by four birds; bands of geometric ornament resembling interlacing dragons on the body and foot; a ring attached to a *taotie* mask in low relief on each side.
Zhou dynasty, eighth-sixth century B.C.
Height 48 cm.
Alan and Simone Hartman Collection, New York

56
Hu with animal head.
Bronze with black and green patina
Zhou dynasty – Period of the Warring States, fifth-third century B.C.
Height 46 cm.
Eskenazi Ltd., London

6 Political occasions: appointments and investitures. 'In the first year of the royal calendar, in the 6th month, at the 3rd quarter of the moon, on the day *yihai*, the king being at Zhou, in the great [hall] of King Mu. [The king] spoke these words: "Hu, I give you the mandate to replace your grandfather and your father as master of the affairs of divination...."'

'In the 3rd month, at the 1st quarter of the moon, on the day *jiaxu [chia-hsü],* the king being at the monument of Kang [K'ang], Xieba [Hsieh Pa] entered to help Kang on his right hand.

57
Very rare goblet with two *taotie* masks in low relief to which rings are attached; on the lower part of the body are small handles shaped like animal masks. There are four small ducks in full relief on top of the cover.
Bronze with a green patina
Late Period of the Warring States, fourth-third century B.C.
Height 20 cm.
Private collection

57

The king ordered: "Run the king's house. I give you red boots and a harness with metal ornaments". Kang bowed, beating his head on the ground, venturing to extol the magnificent generosity of the son of Heaven.

That is why he has had this precious sacrificial vessel made for his distinguished deceased father Libo [Li Po]. May his children and grandchildren use it perpetually for ten thousand years'.

For the officers of the royal household, the royal mandate conferring some distinction or other was usually a confirmation, extension or modification of an earlier mandate bestowed on their fathers or grandfathers by the father or grandfather of the reigning king.

Although fiefs and appointments were not hereditary, they usually remained in the same family. In such cases the king would confirm the title, '…The king said, "Hu, at the time of the ancient kings, the mandate was given to your deceased grandfather and father to serve as officers of the first class, masters of discipline for the two corps of the left and right wings [of the army]. Since I respect the mandates of the ancient kings, I give you the mandate to replace your deceased grandfather and your father as officier of the first class, master of discipline for the two corps of the left and right wings [of the army]. Be respectful from morning to evening and do not neglect my orders. I give you red boots for your service.…" '

7 Official ceremonies. 'Bo Zhifu [Po Chih Fu] made this precious *gui* to receive [or entertain?] the king, between his arrival and his departure'. 'The king reached the temple of Zhou and, under the porch of proclamations, he gave a banquet.…'

8 To commemorate an event, mainly when a present was received from the king. '…The Zhou [king] said: "I give you splendid red sashs, a black robe, belts and skins; I give you a chased spear, red arrows....." Zai [Tsai] bowed, prostrating himself to praise the king's gifts. [Zai] made this precious vessel *dui [gui]* for the deceased Wen. May his sons and grandsons take precious care of it eternally'. 'The king gave cowries to De, 20 strings [of cowries]. Then he [De] made this precious sacrificial vessel'.

9 Vessels as dowry. 'In the 26th year, in the 10th month, at the 1st quarter of the moon, on the day *yimao*. Fan Zhusheng [Fan Chu sheng] had this marriage *hu* made as a gift for the marriage of his first child Meng Feiguai [Meng Fei kuai]. May his children and grandchildren make precious use of it for ever'. 'In the 1st month, at the 1st quarter of the moon, on the day *dinghai [ting-hai]*, the

great guard Cai [Ts'ai] gave this banqueting vessel as a marriage gift to Xu Shujikemu [Hsü Shu Chi K'o Mu].…'

This usage was confined to the seignorial houses and to officers' families. Vessels given as dowry or *yinqi [yin-ch'i]* usually came from the family of the wife's father. Husbands sometimes gave similar vessels to their wives, but examples are comparatively rare. Gifts of this type became commoner under the Eastern Zhou and seem to have become widespread during the Period of the Spring and Autumn Annals.

Vandermeersch believes that texts of this type relate to celebrations of 'the ritual of conferring titles, office, and rewards, of which the inscribed bronze vessel, made at the request of the recipient, formed as it were the letters patent'. These ceremonies usually took place in the royal sanctuaries or, rarely, in some other centre of culture.

By the end of the Western Zhou, the size of the characters and spacing tended to become uniform and the reign and cyclical dates to be named. The Period of the Spring and Autumn Annals marks the beginning of the standardization of writing: all the characters were now of equal size and the spacing was regular. The texts were similar in content to those of the preceding era and show that inscriptions were employed by ministers, officials, officers, etc. and by various states that were subject to the Zhou.

In the Period of the Warring States there was a perceptible change in style and grammar by comparison with the preceding periods. Inscriptions began to appear not only on vessels but also on weapons, metal objects, coins, etc. Writing was no longer uniform, and many variants were found in local or individual styles. Characters of all forms, sizes and appearances occur at this period; they may be composed of large strokes or of delicate ones. The writing known as 'bird's-foot script' was invented at this time.

Although the contents were similar to those of earlier periods, the inscriptions of the Warring States period introduced a new rule which was to become more widespread under the Han: the name of the bronze-founder and the capacity of the vessel (its measurements and volume) were recorded. It is noteworthy, however, that during this troubled period the writing of the state of Qin, the most conservative of the Warring States, was strict and conventional in style. Known as *xiaozhuan [hsiao-chuan]* or 'small seal' style, it represents the first form of standardized writing in China. *Xiaozhuan* style was to be used and enforced by Qin Shihuangdi [Ch'in Shi-huang-ti], the First Emperor and China's unifier.

VOCABULARY

In his work *Jinwenbian [Chin-wen-pien]* (Collection of Inscriptions on Bronzes) Rong Geng [Jung Keng] lists 3000 characters which he breaks down into: 1894 deciphered characters with 13,950 variants and 1199 undeciphered characters with 985 variants.

It is obvious that *jinwen* writing on bronze is not a stable script. There are numerous variations in the characters. A single character may often have over one hundred variants and sometimes as many as 200. It is common for several variants to occur in the same text. Such differences partly account for the complexity of deciphering these inscriptions.

Yet the poverty of the vocabulary led to the creation of non-literary texts that are often monotonous, being composed of stereotyped formulae. Nevertheless, some rare inscriptions are excellent historical documents. These include the Mao Gong *ding* [Mao Kung *ting*] (in the Taipei Museum), the Zha Zhi *pan* [Cha Chih *p'an*], the Kang Hou *gui* [K'ang Hou *kuei*] (formerly in the Malcolm Collection, now in the British Museum), the *guo zhi chi bo pan [kuo chih ch'ih po p'an]* (This 104-character inscription is dated 759 B.C.) and the Nei *gui* [Nei *kuei*] (which mentions the Shang rebellion in the early years of the Zhou dynasty).

The lists show first archaic characters followed by modern characters (except for the 'expressions'), the transcription in *pinyin* and Wade-Giles, finally, if necessary, the English version.

Cyclical Characters

十	甲	*jia [chia]*
㇟	乙	*yi [i]*
丙	丙	*bing [ping]*
口	丁	*ding [ting]*
戈	戊	*wu*
己	巳	*ji [chi]*
甫	庚	*geng [keng]*
辛	辛	*xin [hsin]*
王	壬	*ren [jen]*
癸	癸	*gui [kuei]*

Family

父	父	*fu* = father
母	母	*mu* = mother
祖	祖	*zu* = grandfather
子	子	*zi* = son
孫	孫	*sun* = grandson

Names of Vessels

鼎	鼎	*ding*
壺	壺	*hu*
盂	盂	*you*
殷	殷	*dui [gui]*
爵	爵	*jue*
尊	尊	*jia*
卣	卣	*zun*
監	監	*yu*
盤	盤	*jian*
		pan

Common Words

寶	寶	*bao [pao]* = precious
尊	尊	*zun [tsun]* = epithet with the sense of precious
彝	彝	*yi* = general term for ritual vessels
作	作	*zuo [tso]* = to do
其	其	*qi [ch'i]* = to be able
用	用	*yong [yung]* = to use, usage

月 yue [yüeh] = month

萬 wan = ten thousand

侯 hou = marquis

公 gong [kung] = seignorial duke or chancellor

白 bo [po] = count

年 nian [nien] = year

史 shi [shih] = annalist

媵 ying = vessel given as dowry, marriage (vessel)

王 wang = king

旅 lu = campaign (vessel)

貝 bei = cowries

我 wo = me, I

Expressions

尊 彝 zunyi [tsun-li] = sacrificial vessel, precious vessel

寶 用 baoyong [pao-yung] = precious usage

作 册 zuoce [tso-ts'e] = drafter of charters

旅 彝 luyi = campaign ritual vessel

析 子 孫 xizisun [hsi-tzu-tsun] = mark (emblem of a military corps)

其 子 孫 ⎫
永 寶 用 ⎭ qi zisun yong paoyong [ch'i tzu sun p'ao-yung] = May the sons and grandsons perpetually take precious care of it.

Weapons, Fibulae, Mirrors

3 Longbow fittings *(bang [pang])*. The Shang used immensely powerful longbows. This bronze fitting would have reinforced the bow at the point where the archer placed the arrow. At each end of the bow, where the bowstring was attached, there would have been a jade fitting known as *mi* fixed to the wood of the bow to keep the bowstring taut.

When present, the decoration of a bronze *bang* varies from a simple star to a very complex geometric ornament with *taotie* masks or sometimes animals in relief. Specimens decorated with human figures are very unusual, but those unearthed from tombs M. 40 and M. 20 at Yinxu are so decorated.

4 *Ge [ko]* halberds. Certain types of halberd are nearer in form to daggers than to halberds. They are very variable in shape and were presumably used in hand-to-hand combat to wound the horses of enemy chariots and for sacrifices involving slaughter and decapitation. The blade was fixed to the tip of the haft of a spear and was one of the two Shang weapons of war; the halberdier and the archer were in fact the two fighting men who rode in chariots.

5 *Chu [ch'u]* axes. This term is used to denote two types of axe derived from the halberd. The first, generally considered to be a primitive form of the *ge,* has a V-shaped blade, the length of which is one and a half times

59
Yue axe, with a *taotie* mask and *kui* dragon on the socket.
Bronze with a sea-green patina
Shang dynasty, Anyang period, thirteenth-eleventh century B.C.
Length 20 cm.
Collection of the King of Sweden

60
Part of a *bang* bow inlaid with turquoise terminating in the heads of two stylized horses.
Bronze with a blue and green patina
Shang dynasty, thirteenth-eleventh century B.C.
Private collection

61
Ge halberd, with spiral motifs on the blade; the haft is decorated with openwork.
Bronze with a red and green patina
Late Shang dynasty or early Zhou, eleventh century B.C.
Height 19 cm.; length 26.5 cm.
Private collection

59

60

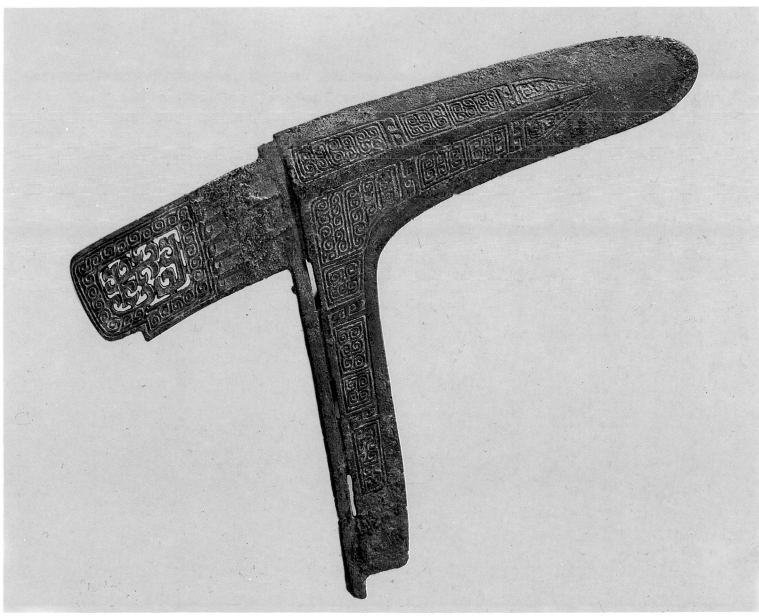

61

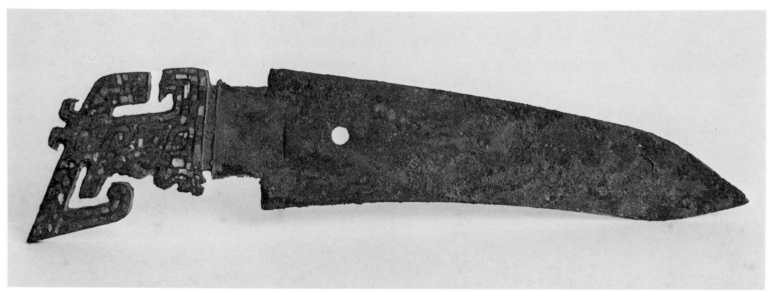

62

63

62
Ge halberd, the haft inlaid with turquoise.
Bronze with green patina
Shang dynasty, Anyang period, thirteenth-eleventh century B.C.
Rijksmuseum, Amsterdam

63
Qi axe with a human mask on the blade.
Bronze with green patina
Shang dynasty, Anyang period, thirteenth-eleventh century B.C.
Height 30.4 cm.; width 35 cm.
Museum für Ostasiatische Kunst, Staatliche Museen Preussischer Kulturbesitz, Berlin

64

the width at the level of the helve. The second variant resembles the *ge*, the only difference being in the socket which is rounded to the shape of the wooden haft into which it fits.

6 *Yue [yüeh]* axes. These small axes with narrow blades and flared outlines terminate in small sockets. The upper part of the blade is decorated either with *kui* dragons or with *taotie* masks, sometimes inlaid with turquoise or very rarely with the figure of an owl.

7 *Qi [ch'i]* axes. Like the *yue* axe, this type of large axe, measuring 30 to 40 centimetres in width, sometimes more, has a flared outline. Very fine specimens are decorated with openwork *taotie* masks or, very rarely, like the one in the Museum für Ostasiatische Kunst, Berlin, with a human face.

8 *Dao [tao]* knives. Some scholars believe that this was the weapon used by the charioteers. The small knives with blades issuing from the mouths of animals or *kui* dragons are Sino-Siberian in style. The very large ones, which may measure over 40 centimetres, are often decorated on the upper part of the blade with dragons or stylized cicadas.

65

64
Warrior's helmet decorated with a stylized *taotie* mask.
Bronze
Excavated from Royal Tomb HPKM 1004, Anyang.
Shang dynasty, Anyang period, thirteenth-eleventh century B.C.
Institute of History and Philology, Academia Sinica, Taipei

65
Profile of the warrior's helmet in Plate 64.

9 Helmets. The royal tombs of Yinxu, in particular tomb HPKM 1004, have revealed the existence of bronze war helmets. They were presumably used by the Shang kings or very high state officials and covered the head from the forehead to the nape of the neck. The little tube at the top suggests that plumes were worn. With a large *taotie* mask sometimes resembling a tiger in front they must have given the warrior a haughty mien indeed.

10 Plates from armour. Chinese archaeologists believe that the small rectangular bronze plates pierced at each corner that have been unearthed in various royal tombs have come from suits of armour.

Weapons under the Zhou were broadly of the same type, but when decorated, the decoration was in the style of the period. It is noteworthy, however, that the sword appears to have become widespread in use in China in about the seventh century B.C.

FIBULAE

Fibulae or clasps form the most representative body of Warring States and Han art. These precious adornments for clothing combine the art of the bronze-founder and that of the goldsmith. Though small at first, in the fifth century B.C., under the Han, fibulae were sometimes of considerable size.

Shapes

Fibulae were utilitarian objects consisting of a body that narrowed at one end to form a hook and a stud at the back by which it was attached to the clothing.

The hook may be plain, or it may represent the head of a dragon, an animal or a monster. In the earliest clasps, which are often undecorated, the hook is well developed and fairly simple. It may also be formed by part of one of the animals or figures of the ornament, such as an extended paw, leg or arm. In other examples the head of the animal may form the hook; whatever its constituents the hook always curves outwards.

The stud may be circular, square or oval and is sometimes decorated with lines, pictograms, motifs in intaglio or relief. Very rarely the stud is formed of an animal head identical with the head of the animal that forms the clasp;

the example in the Musée Cernuschi, Paris, is of this type. The earliest fibulae under the Zhou have very large studs with diameters appreciably greater than the width of the piece. In the Period of the Warring States the stud is usually placed a third of the way along the piece, while under the Han it is often in the middle.

Uses

Fibulae were used for various purposes, such as fastening a belt or securing a cloak: uses presumably varied according to the size of the fibulae. Thus the shortest specimens served a well defined purpose in military equipment: the scabbard of a sword or a quiver were hung from them. The largest were used to adorn clothing, the material and ornament varying according to the rank and perhaps the profession of the owner (witness the clasps decorated with Taoist characters). Some scholars believe that they were used exclusively for funerary purposes, but this does not seem very likely.

Symbolism

Several archaeologists who have studied the ornament of the fibulae have hazarded the guess that they were ritualistic in character. Thus, of the zoomorphic ornament, the dragon, the elephant and the cicada are symbols of resurrection; a bird symbolizes the divinity of the wind; the wheel is an image of the sun. S-shaped motifs, spirals and whorled circles are talismans protecting human beings from spells. Although these few examples suggest that the fibulae had a ritualistic significance, there is no indication that the hypothesis is a reasoned statement and scientifically provable.

66
Cylindrical fibula with geometric ornament.
Gold inlaid with turquoise
Zhou dynasty—end of the Period of the Warring States, fourth century B.C.
Length 18 cm.
Freer Gallery of Art, Washington, D.C.

67
Spatula-shaped fibula with a tiger's head motif in the centre.
Bronze inlaid with silver
Zhou dynasty—Period of the Warring States, fifth-third century B.C.
Length 5.1 cm.
Mr and Mrs Desmond Gure Collection, London

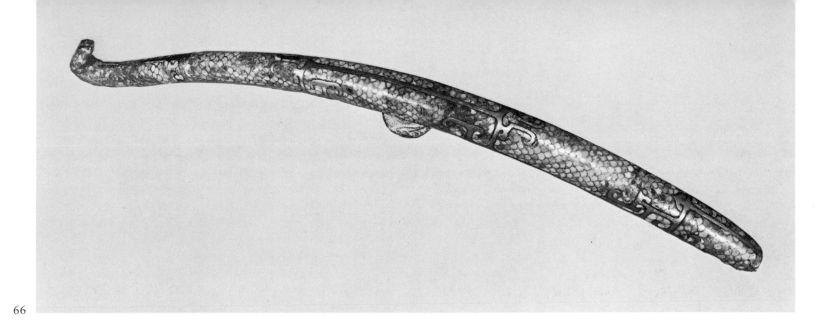

66

67

Zhou Style

Rare during the Zhou period, fibulae became more widespread in China from the third century B.C. This phenomenon was paralleled by changes in dress and in military equipment resulting from assimilation by the Chinese of the techniques of warfare practised by the nomads of the North. The style of fibulae under the Zhou was sober and severe; the enormous stud was larger than a buckle. Clasps were very short, representing zoomorphic heads (a *taotie* mask, the head of an elephant or other animal); hooks and studs were well developed and undecorated.

Style of the Period of the Warring States

Fibulae became widespread in China during this period. The motifs — including spirals, beads, scales, granulation, coils — are rich despite their sobriety and powerful for all their elegance. The influence of nomadic art is manifest in the many animal figures.

Zoomorphic Ornament

There are few large pieces in this category, the length varying between 3 and 9 centimetres. The stud is well developed and occurs at the top of the fibula in the earliest examples and a third of the way along it in the others. Animal figures are usually stylized rather than naturalistic. Among them are some that are ritualistic in character; these include elephants, dragons, cicadas, tortoises, birds, tigers, snakes, hinds and wolves. Although the *taotie* masks and birds are the commonest themes under the Warring States, the influence of the Ordos tribes is apparent in animals in vigorous movement or interlacing.

Non-figurative Ornament

The commonest motif is a triangle accompanied by a coil; lozenges, hooks, granulation, scales, diagonal and horizontal S- and C-curved lines are also frequent. Fibulae of this type may be divided into two groups which are based mainly on the length and shape of the pieces: a) small clasps measuring between 4 and 7 centimetres, with an enlarged base that may be circular, oval or shield-shaped; the stem emerges out of this and tapers to form the hook;

b) large clasps measuring over 10 centimetres in length that are shaped like a slender elongated spatula or the stem of a spoon. The geometric ornament of spirals forms heart-shaped motifs and C-curves. Fibulae measuring 20 centimetres and more are slender and curved and are richly inlaid with gold, turquoises, silver and jade.

Han Style

The Han fibulae as a whole are longer than those of earlier periods, but they may vary in size from 2 centimetres (the rarest examples) to over 20 centimetres. Hooks represent monsters but are neater in style. The stud is nearly always halfway along the length of the clasp.

Zoomorphic Ornament

So great is the variety, due to the immense vogue for fibulae under the Han, that this category may be subdivided into six groups:

1 Plain bird ornament. These clasps are small and must have been used for military equipment. The bird's beak forms the hook while the stud represents the feet. They were sometimes inlaid with glass beads, shell, coloured gemstones and jade.

2 Medium-sized clasps representing animals such as birds, buffalos, horses, bears, monkeys, bats, tortoises, cicadas.

3 Chamfered clasps with animal ornament. They originated in northern China and vary in size between 10 and 15 centimetres. They are broad and heavy in shape with chamfered relief. Strongly influenced by the style of the Ordos tribes, they are decorated with animals from the steppes with sinuous bodies and coiled tails. The ornament is sometimes enriched with spirals and inlay.

68
Large fibula (clasp) inlaid with turquoise, shaped like a bird with wings outspread that is being devoured by four dragons.
Bronze with a green patina
Period of the Warring States, fifth-third century B.C.
Height 19 cm.; width 16 cm.
Tai Collection, New York

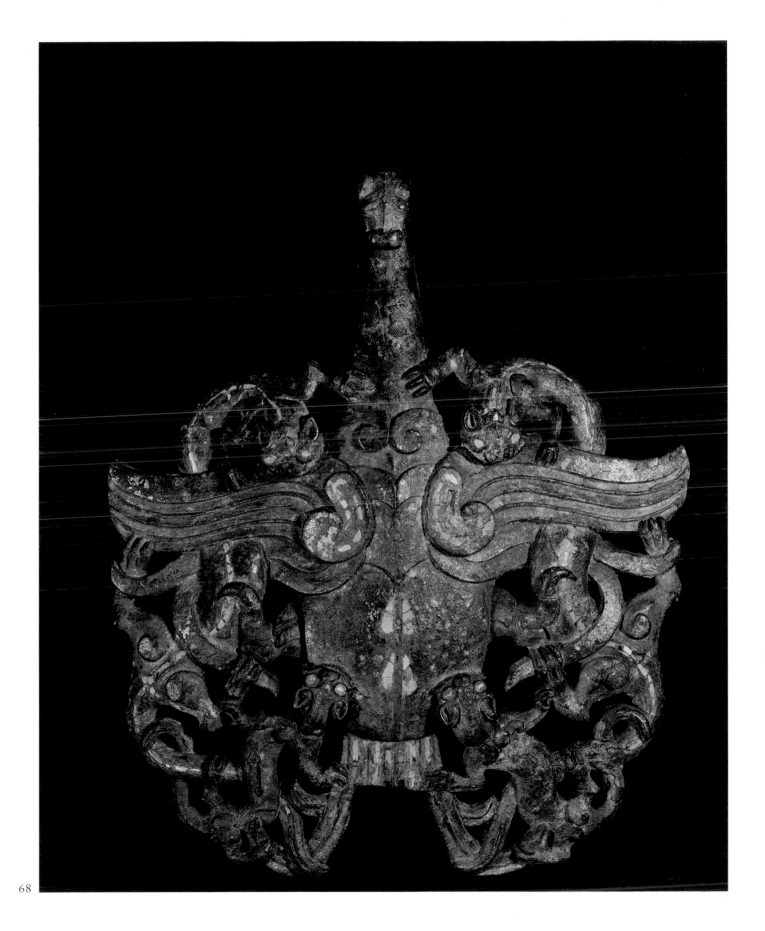

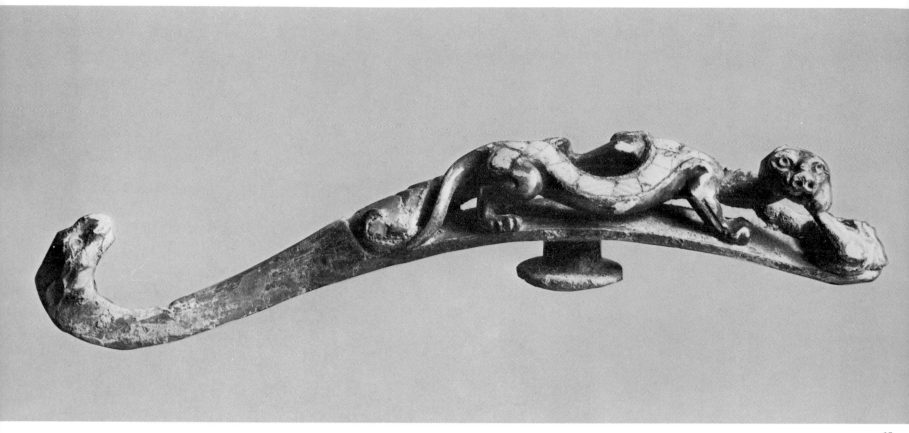

69
Fibula decorated with a dragon in full relief.
Gilded bronze inlaid with turquoise
Han dynasty, second century B.C.
Length 10.2 cm.
British Museum, London

4 Realistic ornament. The animal on these pieces is represented realistically and is modelled in the round.

5 Dragons or serpents. The sole decoration of these fairly long and often slender clasps is a scene of interlacing dragons.

6 Stylized animals. These large pieces with cut-out silhouettes and often with openwork are decorated with stylized animals that retain, however, a more or less realistic appearance.

Anthropomorphic Ornament

Fibulae of this type are comparatively rare and are decorated with representations of human figures. Despite their different attitudes there is a close resemblance between all these small figures. Their average size varies between 6 and 8 centimetres.

Non-figurative Ornament

These large clasps are fairly slender and shaped like elongated, curved spatulas; they are often inlaid with geometric ornament in precious stones or metal. They are usually made of bronze, but specimens exist in iron, jade and solid gold. Fibulae set with turquoises are decorated with lozenges and triangles outlined in gold and silver.

In his *History of Early Chinese Art*, Osvald Sirèn uses another method of classifying Han fibulae. The results are extremely interesting and, indeed, indispensable for those desiring a quick overall view of the work of that period; a résumé of the fourteen groups established by this eminent scholar should therefore prove to be of great value to collectors:

1 The body of the clasp is composed of one or more openwork dragons, i.e. dragons cut out in chamfered

70

silhouette. The dragon forms an S-curve and bites its back. This theme occurs in still more stylized form on small pieces. On large pieces two, and sometimes three, dragons are locked in combat.

2 The body of the clasp is composed of interlacing dragons. Though similar in style to the previous group, these dragons are not in openwork. The clasp is sometimes reduced to one large dragon or a *taotie* head. In such cases the hook is closed by the animal's tail.

3 The body of the clasp is short and wide representing an animal head—either a *taotie* or a feline head—with the body often coiling in spirals. The paws are formed by lateral extensions. The head sometimes has ears and a crest in the form of a bird's tail. The hook is an extension of the muzzle.

4 Relatively large clasps, the bodies of which taper toward the hook. The decoration varies widely but is for the most part based on zoomorphic motifs.

5 Clasps with elongated bodies decorated with fluting and *taotie* masks at the extremities.

70
Fibula inlaid with turquoise, shaped like a sword-bearing warrior who is being eaten by a wild beast.
Gilded bronze
Han dynasty, second century B.C. — second century A.D.
Kunstindustriemuseet, Copenhagen

71

71
Fibula in the shape of two human figures playing musical instruments.
Bronze with dark patina
Length 6.5 cm.
Musée Guimet, Paris

72

72

Fish-shaped fibula.
Bronze inlaid with silver
Han dynasty, second century B.C. — second century A.D.
Brooklyn Museum (gift of the Guennal Collection), New York

6 Bird-shaped clasps. Wings and tails are almost realistic. The bird's neck forms the hook.

7 Spoon-shaped clasps, rather elongated and sometimes slightly convex. They are decorated in relief with running or fighting animals.

8 Small clasps, the body of which represents an entire animal (horse, etc.) or a small human figure in silhouette.

9 Medium-sized clasps with zoomorphic ornament sometimes reflecting a strong Sino-Siberian influence.

10 Clasps inlaid with turquoises or other coloured substances. These pieces are large, with elongated bodies and curved beaks. The decoration of turquoise inlays in the form of geometric figures (lozenges and spirals) is usually heightened by broad gold lines.

11 Clasps of the same type as those of the previous category but decorated in a different way. The ornament is either inlaid with gold or silver on a bronze ground or with a design that stands out against a silver background.

12 Clasps in the form of curved tubes with geometric ornament (lozenges, triangles and spirals) inlaid with silver. Sirèn dates these pieces to the end of the Han and to the Six Dynasties period.

13 Clasps which probably come from the territories of the far northwest. The body is composed of figures of entire animals (dogs, tigers, boars, etc.), either on their own or in groups and locked in combat. The hook is formed of the animal's neck or tail. Solitary figures of animals, sometimes with the head retroverted and in either the couchant or passant posture, were inspired by Scytho-Mongolian art.

14 Clasps, the body of which represents an animal in high relief (often a long-horned ibex running or leaping, or an ibex attacked by a tiger). These pieces are also clearly influenced by Scytho-Mongolian art.

MIRRORS

After the fibulae, mirrors constitute the second most representative category of the art of bronze under the Warring States and the Han. But they had a much longer life: having made their appearance towards the end of the Shang, they were still in production up to the beginning of the twentieth century. So their development — mainly modifications of style, epigraphy and methods of casting — may be studied over a period of nearly 3,200 years. Their shape alone remained constant over the centuries: they were either circular or square and were polished on one side and decorated on the other with ornament that unfolds round a central boss.

Authenticity

Mirrors in collections are numerous, and doubts have arisen in the minds of many scholars as to the authenticity of so many pieces attributed to the Warring States and the Han. Without any doubt some must be Tang, Song, Ming or even later copies. Indeed, we know from ancient writings that Han mirrors were copied by the Sui and by the early Tang. In the same way, mirrors with TLV decoration were the favourite models of the Song, Yuan and Ming founders. All the other shapes dating from the Warring States and the Han were also reproduced. This accounts for the frequent occurrence of pieces in the Han style bearing dates of the Song or Ming periods. However, a fair number of these copies are recognizable from the colour of the bronze, the patina, the style of writing and from those inscriptions that mention the date, place and name of the founder. In the most complicated cases an examination of the weight and the composition of the alloy is essential but does not always prove conclusive in the matter of the authenticity and exact dating of the piece.

In their admirable study of the dating of Chinese bronze mirrors, A. Gutkind Bulling and Isabella Drew express their belief that the crucial test is the relationship between the weight and the diameter of the piece. Indeed, they hold that each period in Chinese history has its own aesthetic which affects this relationship. Thus they have observed that Tang copies of Han bronzes are usually about 25 per cent heavier than the originals.

The use of X-rays and stereomicroscopical examination may also produce interesting results concerning the authenticity of mirrors. For example, the latter method reveals how the piece was cast. Most of the copies were cast by the 'lost wax' *(cire-perdue)* method, whereas the mirrors of the Warring States and the Han were produced by direct casting using moulds.

Metallographical analysis may sometimes produce information on which a date may be hazarded. While ancient mirrors, up to the Tang, have a basis of copper, tin and small quantities of lead, the copies contain in addition zinc, iron and lead in proportions that may amount to 20 per cent. Unfortunately, we have insufficient technical data concerning the numerous mirrors excavated scientifically in China during recent years, and in the present state of knowledge certain bronze mirrors are still impossible to date with precision.

Shang Mirrors

Until 1934, when the royal tombs at Anyang were discovered, the existence of bronze mirrors of the Shang dynasty had never been suspected. As far as scholars knew, they had made their appearance in China around the date of the Warring States (481–221 B.C.). When one intact mirror and part of another were brought to light from tomb HPKM 1001, the known date at which mirrors were used was moved back 500 years. But a slight doubt as to the exactitude of the date remained. It was necessary to wait until May 1976 before this was dispelled with the discovery of four other pieces of the Shang period in the grave of Fu-hao at Xiaotun [Hsaiot'un].

These mirrors are round and very simply decorated. Six protruberances encircling a boss of the same size as themselves is the only decoration on the mirror from tomb HPKM 1001.

73

73
Mirror with six bosses.
Bronze with dark patina
Excavated from Royal Tomb HPKM 1001, Anyang
Shang dynasty, Anyang period, thirteenth-eleventh century B.C.
Institute of History and Philology, Academia Sinica, Taipei

74

74
Mirror decorated with Ts.
Bronze with black patina
Zhou dynasty — Period of the Warring States, fourth-third century B.C.
Diameter 12.5 cm.
Huges Le Gallais Collection, Venice

75
Mirror decorated with human figures.
Bronze with a green and red patina
Han dynasty, second century B.C. — second century A.D.
Diameter 18.4 cm.
Freer Gallery of Art, Washington, D.C.

75

Zhou Mirrors

Despite many excavations in China during recent years, we have no information concerning bronze mirrors dating from the beginning of the Zhou dynasty. With the Periods of the Spring and Autumn Annals and the Warring States, the documents, although still confused, become fairly numerous. Archaeologists call these mirrors pre-Han, pre-Qin [Ch'in], Warring States or Huai style.

In most of these mirrors the reflecting surface is convex, and they have a cylindrical boss decorated with linear motifs. They can be divided into two types according to the form of the rims: a) fairly tall rim, vertical and flat on top. These mirrors are rather thinly cast and are decorated with reptiles or other animals, usually on a ground of spirals or geometric motifs (These mirrors are generally thought to date from the Period of the Spring and Autumn Annals.); b) narrow rim curving inwards towards the centre or everted. The principal motifs — which sometimes have a symbolic meaning, like the character *shan* 山 (mountain) – are always on a ground of *leiwen* or spirals. On the basis of the many similar mirrors unearthed at Changsha [Ch'ang-sha], they may be attributed to the Period of the Warring States.

Of all the pre-Han mirrors those which appear to be the most securely dated are the mirrors of the fifth to the third centuries B.C., i.e. those of the Period of the Warring States. These little mirrors, which are thin and have a very small central boss, have lightly engraved ornament of extreme delicacy. The ornamental motif composed of flowers, dragons, stylized birds, geometric motifs (lozenges, frets, zigzags, etc.) is always set off by a ground of spirals, hooks or meanders.

Han Mirrors

Mirrors were very much in vogue from the Period of the Warring States onwards and reached their highest point under the Han, when, with the fibulae, they became the most representative manifestation of the art of bronze in China.

Although similar in shape to the earlier pieces, there was an important variation in the appearance of the Han mirror in that the central boss was now hemispherical instead of cylindrical, as it had been under the Warring States.

Former Han or Xi Han [Hsi Han]

The reflecting surface of the mirrors of this period is flat; they are usually decorated with birds or animals drawn in fine lines to form an interlaced motif. A square or circular frame surrounds the central boss and usually contains an inscription composed of ten or twelve characters. These mirrors may be classified in eight groups according to the type of decoration or the inscription:

1 Shouzhou [Shou-chou] Mirrors   The name refers to a region of the Huai valley where a large number of these mirrors were unearthed. They are very thin and are decorated with interlacing dragons (*panchi [p'an-ch'ih]*), coils and triangles on a ground of *leiwen*. This type seems to make its appearance from the second half of the third century B.C., and although Karlgren considers these mirrors to be pre-Han, it would appear that they were made continuously until the first century B.C.

2 TLV Mirrors   The most important innovation under the Han was certainly the practice of inscribing mirrors. The second was the introduction of TLV motifs into the decoration. This type — the European term for the three letters has been adopted by Chinese archaeologists — was to be the commonest from the second century B.C. onwards.

3 Caoye [Ts'ao-yeh] Mirrors   This term denotes mirrors the decorative motif of which is composed of a central square containing an inscription and decorated with stylized flowers in each corner.

4 Mirrors with Simple Spirals   There are two variants of this type, which was particularly popular between 130 and 80 B.C. The first is decorated with S-shaped spirals. The decoration of the second is composed of four spirals made up of three open Cs placed alternately. The motif is derived from the dragon from which the zoomorphic element has disappeared.

5 Siru Sihui [Ssu-ju Ssu-hui] Mirrors   These mirrors were very popular at the end of the first century B.C. The decoration consists of four nipples and four dragons, each occupying about a quarter of the diameter of the mirror. The central boss is framed by a small square.

6 Mirrors Decorated with Characters   Some mirrors dating from the middle of the Former Han, mainly about the first century B.C., bear inscriptions as their sole decoration. Some types are known as *erzi [erh-tzu]* from the character *er* which appears between each character of the inscription; others decorated with several concentric zones containing inscriptions are called *Zhongjuan [Chung chü-an]*; others again are known as *neixiang lianhu [nei hsiang lien-hu]*. These are also decorated with characters but the centre is in the form of arcs of a circle.

7 Qingbai [Ch'ing-pai] *or* Mingguang [Ming-kuang] *Mirrors* These names have been given to mirrors on which the characters *qingbai* and *mingguang* appear. The ornament consists of bands decorated with spirals or circles and joined by straight or diagonal lines. Most of these mirrors have been unearthed in Shandong [Shan-tung] province.

8 Bairu [Pai-ju] *or 'Hundred-nipple' Mirrors* These mirrors are also known as *xingyun [hsing-yün]* (stars and clouds) or *lianzhou [lien-chou]* (string of beads). They seem to have made their appearance at the end of the Former Han during the reign of Wudi; they are decorated with nipples linked by curving lines or clouds. These motifs are cosmic symbols and, according to ancient texts, represent the constellations of the solar system.

Mirrors of the Reign of Wang Mang

The period of Wang Mang's usurpation of power is marked by the development of a decorative technique based on sectioning the surface of the mirror. The base of the hemispherical boss is decorated with four flower-petals. The principal motif, supplemented by elements resembling the letters T, L and V, is made up of four mythical animals: the unicorn, phoenix, tortoise and dragon, each occupying a quarter of the surface.

Mirrors of the Later Han *or* Dong Han [Tung Han]

The decoration of mirrors changed under the Later Han, as did the phrasing of inscriptions. The commonest and presumably the most popular mirrors are decorated in high relief with human figures, figures of anthropo-morphic divinities and immortals. The repertoire of animal figures is larger: tigers, horses with or without chariots, rhinoceros, monkeys, birds, etc. make frequent appearances. A decorated zone composed of arabesques, meanders, coils and clouds has been introduced into the borders of the mirrors.

Important modifications of style affected the TLV mirrors, which enjoyed a great vogue under the Later Han: dragons and arabesques were replaced by figures of moving animals, strange monsters and anthropomorphic immortals. All these figures are represented in movement, hunting, fighting or in animated conversation.

Two forces determined the development of Han mirrors: the continuity of religious concepts and the cosmic symbol of the universe as seen by the Chinese. While the Former Han executed elegant dragons and arabesques, the art of the Later Han, dominated as it is by figures of deities, shows a degree of stagnation. The new decorative motif reflects the intellectual changes wrought by the influence of Buddhism on Confucianism and Taoism.

Tang Mirrors

The great advances of Chinese art and culture under the Tang had their counterpart in the decoration and manufacture of mirrors. There was now an immense variety of shapes, from circular to square, flower-shaped or multilobed and handled. The central boss might be hemispherical or zoomorphic. Only the reflecting surface remained unchanged, that is to say convex, as in previous periods. The composition of the alloy was also different, the high silver and tin content lending the mirrors a silvery-white appearance.

Ornament was sometimes inlaid with gold, silver or mother-of-pearl and drew its inspiration from: nature, i.e. plants, animals, fish, ducks, etc.; folklore, e.g. Wang Zhi [Wang Chih] watching the immortals playing a game, the five sons of Yan Shan who became high officials, Chang Ou flying to the moon, etc.; and mythology or religion, with Taoist cosmology as the principal element, e.g. the planets, the eight trigrams, immortals seated on dragons, etc.

Though still in the Tang tradition, mirrors became much more baroque under the Song [Sung] and the workmanship was poorer. The inscriptions often permit such mirrors to be attributed to that period. With the spread of glass mirrors under the Yuan and the Ming, bronze mirrors gradually disappeared.

76

76
Lobed mirror decorated with leafy arabesques and flying birds.
Bronze backed with gold
Tang dynasty, eighth-ninth century
Diameter 15 cm.
Freer Gallery of Art, Washington, D.C.

Inscriptions on Mirrors

As has already been noted, the Han broke new ground when they introduced inscriptions into the decoration of their bronze mirrors. The content of the inscriptions varied from period to period. Thus under the Former Han the phrases were of the type: 'The inside [of this mirror] is made of pure material so that its reflections are bright; its light resembles the sun and moon with no difference'. 'As we look at the brilliance of the sun we shall not be able to forget ourselves'. 'As we look at the light of the sun, everybody is very greatly illuminated'.

During the reign of Wang Mang, inscriptions frequently mentioned the name of the new dynasty, Xin [Hsin].

Under the Later Han, inscriptions sometimes name Dong Wangcong [Tung Wang Ts'ung] (Royal Duke of the West), Xi Wangmu [Hsi Wang Mu] (Queen-Mother of the East) or the places where the mirrors were made, like Shangfang, Danyang [Tang-yang], Dongliang [Tung-liang], etc. The content of the texts themselves also changed and was now of the type: 'May you have sons and grandsons [for] a long time'. 'The red bird and the black guardian symbolize the *yin* and the *yang*'. 'The dragon on the right, the tiger on the left watch over the four directions'. 'Every day may you have joy, wine and things to eat in abundance. May you rejoice in being sheltered from all vexations'. 'The imperial administration has made this mirror with great skill. There are immortals on it who do not know old age. When they are thirsty they drink at the jade springs'. 'May your family always be honoured and rich'.

Under the Tang and the Song, inscriptions, which are in a very different style of calligraphy; they express wishes for happiness and prosperity for the persons who made or owned the mirrors: 'Riches, honours and long life'. 'Joyous prosperity, descendants of honour'. 'Double felicity of happiness and long life'. 'Perfect fortune by the dragon and the phoenix'. 'The venerable family X of [name of place] have cast this mirror of bronze without equal'.

77

77
Eight-lobed mirror with embossed pattern of mythical animals in vine-scrolls.
Bronze backed with silver foil
Tang dynasty, eighth-tenth century
Diameter 19.4 cm.
Hakutsuru Museum, Kobe

Relation in Mirrors of Weight to Diameter

(after A. Gutkind Bulling and I. Drew)

Diameter cm.	Warring States Mid Western Han	Late Western Han	Wang Mang/ TLV Mirrors	Wang Mang and Eastern Han	Eastern Han/ Quatrefoil Mirrors	Late Eastern Han/ Dragon, Tiger Mirrors	Sui and Tang/ Lobed Mirrors	Tang/ Grapevine, Lion Mirrors
8.0	40–67	77–93	–	92–112	–	–	–	168–206
8.5	45–77	88–108	–	105–129	85–105	126–154	–	198–242
9.0	50–87	100–122	–	121–147	96–118	145–177	–	231–283
9.5	57–98	113–139	–	138–168	108–132	166–202	–	267–327
10.0	64–109	128–156	165–210	155–189	121–147	188–230	207–253	307–375
10.5	71–120	142–174	190–235	174–212	133–163	212–260	234–286	349–427
11.0	79–133	158–194	220–270	194–238	147–179	238–290	263–321	396–484
11.5	90–146	175–215	260–300	216–264	161–197	265–325	294–360	445–545
12.0	115–161	194–238	299–337	239–291	176–216	294–360	328–400	499–611
12.5	144–176	213–261	322–362	263–321	193–235	326–398	363–443	557–681
13.0	156–190	233–285	357–385	289–353	209–255	359–439	400–488	618–756
13.5	169–207	254–310	392–432	316–386	227–277	393–481	439–537	684–836
14.0	183–223	276–338	426–474	345–421	245–299	430–526	481–589	754–922
14.5	197–241	300–366	463–519	374–458	264–322	470–574	526–642	828–1012
15.0	212–260	324–396	507–559	406–496	284–348	510–524	572–700	907–1109
15.5	228–278	349–427	548–606	438–536	304–372	553–677	621–759	991–1211
16.0	243–297	375–459	592–654	473–577	326–398	599–731	673–823	1078–1318
16.5	260–318	403–493	638–706	508–622	348–426	–	726–888	1171–1431
17.0	277–339	432–528	679–765	545–667	–	–	783–957	1268–1550
17.5	295–361	462–564	713–837	585–715	–	–	842–1030	1371–1625
18.0	313–383	492–602	771–896	625–763	–	–	904–1104	–
18.5	332–406	525–641	816–958	667–815	–	–	968–1184	–
19.0	352–430	558–682	870–1022	710–868	–	–	1035–1265	–
19.5	365–447	581–709	949–1113	741–905	–	–	1104–1340	–
20.0	392–480	627–767	965–1179	802–980	–	–	1177–1439	–

scrolls, sixth to fifth century B.C.

leaves on a background of scrolls, sixth to fifth century B.C.

inclined T motifs, third century B.C.

dragons on a background of intertwined T's, third century B.C.

star-shaped motif, third century B.C.

intertwined stylized dragons, third to second century B.C.

intertwined dragons and zigzags, third to second century B.C.

TLV motifs and very stylized dragons, second century B.C.

star motif with four dragons, first century B.C.

TLV motifs, tigers; dragon band around the rim, first century B.C. to first century A.D.

multifoil mirror: intertwined flowers and birds, Tang dynasty

multifoil mirror: scroll decoration; border of petals, Tang dynasty

The Han Dynasty

Chronology of the Han Dynasty

Western or Former Han:	206 B.C.—A.D. 8
Gaozu [Kao Tsu]	206-195 B.C.
Huidi [Hui-ti]	195-88 B.C.
Regency of the Empress Lu [wife of Gaozu]	188-80 B.C.
Wendi [Wen-ti]	180-57 B.C.
Jingdi [Ching-ti]	156-41 B.C.
Wudi [Wu-ti]	141-87 B.C.
Zhaodi [Chao-ti]	87-74 B.C.
Xuandi [Hsüan-ti]	74-48 B.C.
Yuandi [Yüan-ti]	48-33 B.C.
Chengdi [Ch'eng-ti]	33-7 B.C.
Aidi [Ai-ti]	7-1 B.C.
Pingdi [P'ing-ti]	1 B.C.—A.D. 8
Usurpation of Wang Mang	9-22
Eastern or Later Han:	A.D. 25-220
Guang Wudi [Kuang Wu-ti]	25-58
Mingdi [Ming-ti]	58-76
Zhangdi [Chang-ti]	76-89
Hedi [Ho-ti]	89-106
Shangdi [Shang-ti]	106-7
Andi [An-ti]	107-26
Shundi [Shun-ti]	126-45
Chongdi [Ch'ung-ti]	145-6
Zhidi [Chih-ti]	146-7
Huandi [Huan-ti]	147-68
Lingdi [Ling-ti]	168-89
Xiandi [Hsien-ti]	189-220

Once the China of the eighteen provinces had been unified, Qin Shihuangdi [Ch'in Shih-huang-ti], (the 'First August Emperor' of the short-lived Qin [Ch'in] dynasty 249–07 B.C.) set to work to standardize weights, measures and writing and to construct an ambitious road-system. This great administrator, who seems to have been a detestable and cruel monarch, retains the sad distinction of having ordered the burning of all the classical books of China. His dynasty did not long outlive him.

In 206 B.C. after a period of confusion, a small landowner named Liu Bang [Liu Pang] founded the new dynasty of the Han. This dynasty, which ruled China until A.D. 220, apart from a short interruption when Wang Mang usurped power (A.D. 9–22), was an extremely refined civilization, achieving territorial expansion and commercial exchanges.

THE FORMER HAN (QIAN IIAN [CH'IEN HAN) OR WESTERN HAN (XI HAN [HSI HAN]: 206 B.C.—A.D. 8

Liu Bang became emperor, assumed the name of Gaozu [Kao-tsu]) and instituted a policy of large-scale public works, most of them in the fields of strategy and economics. He was an able and prudent man, who understood the art of government, and surrounded himself with numerous officials and advisors. With the aim of anticipating possible rebellion and movements of independence against the imperial power, Gaozu removed his former comrades in arms whom he had placed at the head of fiefs and substituted members of his family. After Gaozu's death, Huidi [Hui-ti] strengthened the central power by reinstating the scholars, and in the year 191 B.C., he annulled the edict of Shihuangdi who had ordered the destruction of all classical books and the death penalty for all who flouted that law.

The policies of Gaozu and Huidi were dominated and directed by the problem of the invading horsemen or nomads of the steppes, the Xiongnu [Hsiung-nu]. The Xiongnu, including Mongolian, Turkish, Tungus and other tribes, plundered the north of the country incessantly. The early Han emperors exercised a policy of appeasement towards them, which was known as *heqin [ho-ch'in]*, i.e. 'peace and friendship'. This consisted of regular consignments of expensive gifts of silks, alcohol, copper coins, rice, etc. Sometimes too a Chinese princess would be given to their chief in marriage.

With the succession to the throne of Wendi [Wen-ti] (180–57 B.C.), one of the greatest of the Han emperors, the Chinese attitude towards the Xiongnu changed. On the advice of his two principal counsellors Jia Yi [Chia Yi] (200–168 B.C.) and Chao Cuo [Ch'ao Ts'o] (?–154 B.C.), who criticized the policy of appeasement on the grounds that it increased the power and wealth of the barbarians, and in the face of growing demands from the Xiongnu, Emperor Wendi resolved upon a radical change

78
Crouching human figure.
Bronze with green patina
Former Han dynasty, second—first century B.C.
Height 8.5 cm.
Musées Royaux d'Art et d'Histoire, Brussels

78

79

79
Niaoxizun or *zun* in the shape of a duck.
Bronze with green patina
Han dynasty, second century B.C. — second century A.D. (or possibly end of the Period of the Warring States)
Brooklyn Museum (gift of the Guennal Collection), New York

80
Fang hu or square *hu* with a *taotie* mask in low relief on each side to which rings are fixed; the neck and foot are decorated with triangular motifs.
Gilded bronze
Former (Western) Han dynasty, 206 B.C. — A.D. 8
Height 45 cm.
Private collection

of direction: he decided to defend the northern approaches and to resume the policy of expansion. During the reign of Wudi [Wu-ti], the 'warrior emperor' (141–87 B.C.), the new strategy was to result in the first great victorious offensives against the Xiongnu (between 127 and 119 B.C.). For this purpose, expeditionary forces numbering over 10,000 men (horsemen and foot-soldiers), with entirely new weaponry designed for rapid movement, were thrown into battle against the terrible nomadic hordes, restoring Chinese suzerainty in the north of the country as far as the Great Wall.

Thus the ground was prepared for the beginning of the great period of expansion, conquest and commercial exchange. This took concrete form in the annexation of the kingdoms of southern China, expansion south of the Yangtze River, invasion of the Nan Yue [Nan Yüeh], of the region of Thanh-hoa, of northern Annam, etc. as well as penetration into the interior of Mongolia, to the south

82

81
Dilianghu, a tripod vessel of bulbous shape; the cover is surmounted
by three rings or stylized birds.
Bronze with green patina
Dynasty of the Former Han, second–first century B.C.
Height (with chain) 23.5 cm.
Eskenazi Ltd., London

82
Chariot with sunshade and horse.
Bronze with a green patina
From the Tomb of General Chang at Leitai, Gansu province (October
1969)
Later (Eastern) Han dynasty, first-second century A.D.
Height (chariot) 43.5 cm.; (horse) 40 cm.

of Manchuria and into northern Korea. The causes and
consequences of this territorial expansion and these
military campaigns were many, but they were also closely
interconnected. They included an economic boom due to
commerce and trade, which considerably increased the
wealth of the merchants, those in high positions and the

Empire; the discovery of great commercial routes linking
southern China with South-east Asia and Central Asia
with India and Iran; the opening up of the Silk Road via
Central Asia and the oases; progress in the metallurgy of
iron, enabling the Chinese to make new weapons (the
sword and crossbow, with consequent reformation of the

84

83
Jing, a cylindrical vessel on three legs with a cover and a chain from which to hang it.
Bronze with dark green patina
Han dynasty, second century B.C. — second century A.D.
Height 19 cm.
British Museum, London

84
Lian with three legs in the shape of bears; the vessel is decorated with two *taotie* masks in low relief and a pattern of meanders, clouds and spirals.
Gilded bronze with encrusted green patina
Han dynasty, second century B.C. — second century A.D.
Victoria and Albert Museum, London

HAN BRONZES

Fibulae and mirrors are the two most representative branches of the art of bronze under the Han. Since they have been discussed in the preceding chapter, the following pages will deal only with the various types of ritual vessels. Because most of these vessels were either made for the lower aristocracy, the feudal lords or rich merchants or were for everyday use, they were produced in considerably larger numbers than before. The immediate result of the new conditions was casting of poor quality, due to a shortage of copper, which was made good by a higher lead content in the alloy (sometimes amounting to over 20 per cent). The second consequence was purely iconographic: the repertoire of ornament dwindled; relief disappeared. Pieces were very often entirely undecorated, in some cases the sole decoration was a *taotie* mask at the points where handles or rings were attached.

Side by side with these everyday articles of medium, not to say mediocre, quality went a splendour that beggars description in the pieces reserved for the 'great ones' of the Empire. Recent discoveries in China have yielded excellent examples. For instance, the tomb of Princess Dou Wan [Tu Wan] excavated in 1968 at Mancheng contained a pair of recumbent bronze leopards inlaid with gold and silver, with a precious stone in each eye-socket, lending a strange red glow to their gaze, and a gilded bronze lamp with shade representing a young woman herself carrying a lamp.

This richness is also found in *inlaid* pieces. Here the ornament consists of inlays of precious metals (gold and silver), enamel and turquoises. The bronze is sometimes engraved or incised, and the recesses are filled with fine gold or silver wire. These bands of scrollwork, spirals, coils occur in relief on pottery urns. Inlay by this technique is, however, stiffer and less graceful than that of the Period of the Warring States.

The same ostentation is seen on *painted* pieces. Pieces of this type are extremely rare, but a bottle preserved in the Musée Cernuschi, Paris, a *ding* in the Osaka Museum and two *lians* that are painted on the insides only, one in the Victoria and Albert Museum and one in the Osaka Museum, may be cited. These pieces are quite often dated to the Former Han. The entire surface of the vessel is covered by a polychrome (red, green and blue) display of triangles, spirals or coils. In rare instances, as in the case of the *lians* already mentioned, the inside of the cover is decorated with a phoenix entirely painted in green on a red ground heightened by polychrome spirals.

Finally this richness is present in *gilded* pieces. The vessels of gilded bronze are probably the most remarkable as well as the most spectacular examples of Han art. A few rare objects of this type with their original gilding have come down to us. One of the finest of these,

85
Hu with two *taotie* masks in low relief.
Bronze with green patina
Han dynasty, second century B.C. — second century A.D.
Musée Cernuschi, Paris

86
Bottle-shaped *hu* with triangular motifs and scrolling foliage.
Gilded bronze with silver niello
Han dynasty, second century B.C. — second century A.D.
Height 36 cm.
85 Musée Guimet, Paris

87

87
Bianhu, an oval-shaped variant of the *hu*.
Bronze with dark patina
Han dynasty, second century B.C. — second century A.D.
Musée Cernuschi, Paris

88
Boshanlu 'hill' censer inlaid with gold, silver, turquoise and carnelian
Gilded bronze
Han dynasty, second century B.C. — second century A.D.
Height 17.9 cm.
Freer Gallery of Art, Washington, D.C.

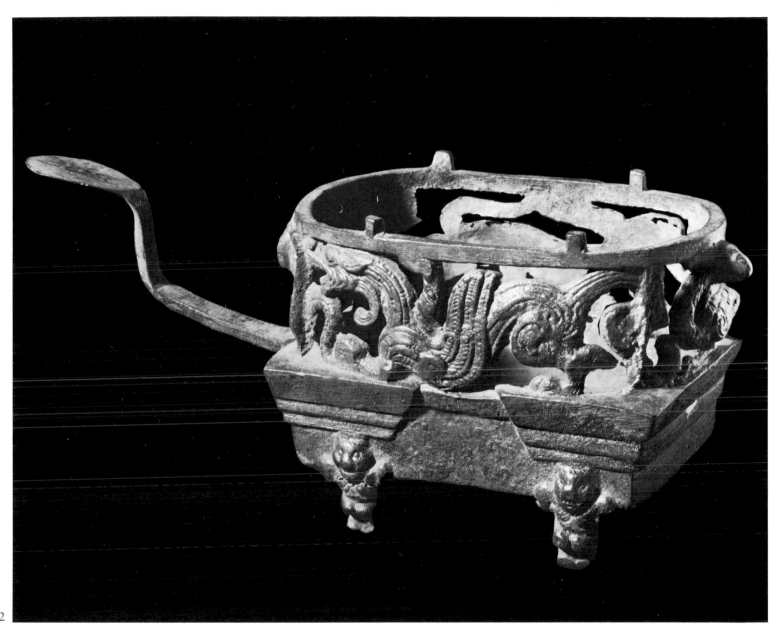

92

91
Boshanlu, a censer with openwork cup supported by a dragon in full relief.
Bronze with green patina
Han dynasty, second century B.C. — second century A.D.
Musée Guimet, Paris

92
Xishen wenlu, a kind of brazier on four legs in the shape of human figures; the upper part of the body is decorated with four animals (tiger, tortoise, duck and dragon).
Bronze with dark patina
Han dynasty, second century B.C. — second century A.D.
Musée Cernuschi, Paris

Animal Style

Animals, especially those in the iconographic repertoire pertaining to handles, are not purely Chinese in origin. They are, in fact, related to the animal style of Iran, the steppes and the Ordos of Inner Mongolia. These foreign influences, assimilated and reproduced in a purely Chinese style, were introduced into China by the Mongolian nomadic tribes, the Xiongnu.

The following figures show distinct affinities with the art of the Ordos and even with Scythian art: the stag and ibex couchant, with legs folded under the body and

93

stylized antlers or horns; the long-eared onager; lions and tigers in curving or half-coiled postures; confronted winged horses; griffins and trampling lions and tigers attacking or devouring a stag, an onager or a mule.

However, the real Han contribution to the dawning age of sculpture was its animal figures as detached statuary. The most characteristic and remarkable examples are gilded bronze bears, seated on their haunches or standing on their hind-legs, (the most illustrious example was in the former D. David-Weill Collection); horses, rhinoceros, tigers and leopards, human figures, usually small crouching, static figures, and the very rare figures in which some semblance of movement is attempted (like the four pieces in the Musées Royaux d'Art et d'Histoire).

93
Jiadou, a pouring vessel on three legs with a spout in the shape of a winged creature.
Bronze with green patina
Han dynasty, second century B.C. — second century A.D.
Musée Cernuschi, Paris

126

94

94
Handle shaped like a snow leopard (?)
Bronze, dark patina with lighter patches
Han dynasty, second century B.C. — second century A.D.
Height 14.5 cm.; width 27 cm.
Sold C. Boisgirard and A. de Heeckeren, Paris, 22 April 1980

95
Small rhinoceros.
Bronze with dark patina
Han dynasty, second century B.C. — second century A.D.
Height 3 cm.; length 7 cm.
Musées Royaux d'Art et d'Histoire, Brussels

95

Bronzes of the Former Han

The problem of classifying the bronze vessels of the Former Han and the Later Han has by no means been solved. However, new archaeological discoveries have made a provisional classification possible, though it may yet need to be modified.

Under the Former Han the decorative tradition of the bronzes of the Warring States underwent considerable changes. Although the technique of inlay persisted — though it was far stiffer in execution — motifs became more secular and began to portray mundane things, such as everyday life and hunting scenes. Ornament was composed of snakes and stylized clouds, animals (tigers, gazelles, camels, elephants, bears, birds), stylized animals in relief or rudimentary landscape or mountain scenes.

One special variety is seen as typical of this period and is represented by bronze vessels with engraved decoration. The objects belonging to this group have been excavated mainly in Luoyang, Xi'an [Sian], North Korea, North Vietnam, at Thanh-hoa and in the region of Changsha [Ch'ang-sha]. The ornament consists of geometric motifs (lozenges and triangles), landscapes and animals shown in profile or in movement (dragons, tigers, birds). In very rare cases the animals' bodies are represented in high relief. When the size or shape of the vessel permits — the *lian* or the *hu* — the different scenes are separated by bands of geometric ornament.

The discovery in June 1968 of the tombs of Prince Liu Sheng (155-13 B.C.) and Princess Du Wan (141-87 B.C.) provides an example of the technical development and abounding creativity of the Former Han when executing articles for the great ones of the Empire. Prince Liu Sheng was the brother of Emperor Wudi and was appointed king of Zhongshan [Chung-shan], a territory which fronted the Xiongnu hordes and comprised fourteen provinces and a population of some 600,000. All the grandeur and luxury of the Former Han are reflected in the sumptuousness, quality and immense variety of the objects unearthed in these two tombs. They include vessels of the *hu* type with a cover (called *zhong [chung]*),

96
Ram-shaped lamp with tip-up back.
Bronze with green patina
Han dynasty, second century B.C. — second century A.D.
Height 13 cm.
Musée Guimet, Paris

97
Another view of Plate 96.

96

97

98

decorated with cloud scrolls or bird script *(niao-zhuan)*, the whole inlaid with gold and silver; a gilded bronze pillow inlaid with jade and a double cup of bronze inlaid with jade and representing a phoenix.

Bronzes of the Later Han

Art under the Later Han became increasingly permeated with humanism and secular preoccupations, testifying to a certain sense of ease and wellbeing. The vessels are quite similar to those of the earlier period, and only a few variants suggest that they date from the Later Han.

Animal style, however, is the most remarkable manifestation of this period, as is proved by the burial of a general of the end of the Han (first to third century A.D.)

98
Chariot with horse.
Bronze
From the Tomb of General Chang at Leitai, Gansu province (October 1969)
Later (Eastern) Han dynasty, first-second century A.D.
Height (horse) 40 cm.

excavated in October 1969 at Leitai [Lei-t'ai]. This tomb, which was probably contemporary with the reigns of the Emperors Lingdi [Ling-ti] and Xiandi [Hsien-ti], the two last Han emperors, contained bronze figures of a cavalry detachment comprising thirty-nine horses, fourteen wheeled vehicles, waggons and chariots with their armed passengers, seventeen horsemen holding spears and

100

99
Fang or *fang hu,* a square version of the *hu,* with two *taotie* masks in low relief to which two rings are attached.
Bronze with green patina
Han dynasty, second century B.C. — second century A.D.
Musée Cernuschi, Paris

100
Small cooking-stove on four legs. Though common in pottery, it is comparatively rare in bronze.
Bronze with green patina
Han dynasty, second century B.C. — second century A.D.
Musée Cernuschi, Paris

101

102

halberds and twenty servants. The style of these figures is similar to that of the Former Han but differs from it in certain new movements. For example, the caracoling gait of the charger was to become a very familiar feature of Tang pottery figures from the eighth century onwards, and the flying or galloping horse balancing by the right hind hoof on a darting swallow was also introduced.

Other distinguishing examples of the art of bronze under the Later Han are the gilded vessels decorated with scrollwork, peacocks, dragons, tortoises, tigers, birds and clouds. There is probably little possibility of dating the ordinary, often undecorated, pieces to any precise phase of the dynasty.

With the fall of the Han dynasty, the art of bronze ritual vessels virtually disappeared from China. But since the Chinese regarded bronze as a precious metal, bronze-working was not entirely abandoned; it was revived in the Six Dynasties and used to create Buddhist images.

104–107
Flying *apsaras*, each holding a different musical instrument.
Gilded bronze
Northern Wei dynasty, sixth century
Height (104) 11.1 cm. (105) 11.4 cm. (106) 9.3 cm. (107) 8.3 cm.
Fogg Art Museum (Grenville L. Winthrop Bequest), Cambridge, Mass.

dated Buddha figure is the seated one in gilded bronze in the Avery Brundage Collection in San Francisco (Pl. 111). The inscription records that it was cast at the end of the Zhao dynasty in the year 338. The Buddha is seated cross-legged on a low dais with hands folded in the *dhyana-mudra* (gesture of meditation) posture. The

drapery and undulating head-dress proclaim the influence of Greco-Buddhist art from Gandhāra, Pakistan, modified to a considerable extent by the art of Asia.

By the fifth century, Buddhist bronzes had become very common. There are extremely fine examples in the Metropolitan Museum of Art, the Nezu Museum, the British Museum, the Seattle Art Museum and elsewhere. These objects played an important part in introducing Buddhism into China, for, being small, they were used for private devotion and were placed on the family altar. Unfortunately a great many bronzes were destroyed during the persecutions of Buddhism in China, mainly in 714 and 845.

Together with the statues and gilded bronzes, Buddhism brought to the Chinese world architectural techniques that were native to the Indo-Iranian border and to India. One result of this was the construction of grottos hollowed out of rock. The earliest of these are known as the Caves of the Thousand Buddhas (Qian-fodong [Ch'ien fo-tung], near Dunhuang [Tun-huang], work on which began in c. A.D. 366.

Between the fifth and the eighth centuries cave art with rock statuary involving statues of colossal size abounded in northern China. The finest and most important groups, put up at the instance of emperors and witnesses of religious fervour, are the caves of Yungang [Yün-kang],

west of Datong [Ta-t'ung], where work began in 489 (The tallest statues rise to between 40 and 50 metres in height.); the caves of Longmen [Lung-men], south of Luoyang (the new capital of the Wei from the end of the fifth century); here work was interrupted during the sixth and seventh centuries. The black stone of Longmen can be worked with great delicacy, a fact which enabled the sculptors to become masters of their art. The art of Longmen is characterized by its careful detail and the extremely high quality of the backgrounds, which occasionally become as important as the main subject. Finally, there are the caves of Maijishan [Mai-chi Shan] near Tianhui [T'ien-hui] in Gansu [Kansu] province.

108

109

108
Small statue of a female donor holding a lotus branch in her hand.
Gilded bronze
Northern Wei dynasty
Height 14.1 cm.
Fogg Art Museum (Grenville L. Winthrop Bequest) Cambridge, Mass.

109
Guardian standing on the back of a lion.
Gilded bronze
Northern Wei dynasty, beginning of the sixth century
Height 7.3 cm. (enlarged)
Eskenazi Ltd., London

TECHNIQUE

There are many literary and historical references to show that bronze was the favourite material for making Buddhist objects under the Southern dynasties. But it did not begin to rival stone in the north of the country until the sixth century. Although it was regarded as the most precious material during the earliest centuries of Buddhism in China, bronze was sometimes replaced by wood, lacquer, clay, ivory and, more rarely, by gold, silver or precious stones.

Composition

Under the Six Dynasties, bronze was usually an alloy of copper and tin in the proportion of nine to one. But certain pieces were executed in pure copper. In other cases tin was replaced by zinc.

When finished the images were nearly always gilded. The process consisted in applying gold leaf with the aid of mercury, which made the gold adhere better and ensured that the composition was uniform. Certain scholars believe that gilding suggested the radiance or 'sun-like luminosity' of the visible body of the god.

Casting

Two methods were used in making Buddhist bronzes, the *cire-perdue* (lost-wax) process and embossing.

The *cire-perdue* technique consists of four processes: a model is made of clay or wood; the matrix is coated with several layers of liquid wax; this is allowed to harden and then is itself coated with heat-proof clay, to form the outside of the mould. As the whole thing is heated, the wax melts and escapes through the outlets provided for the purpose. Liquid bronze is then poured into the mould and replaces the wax. When it has cooled, the mould is opened; the piece is polished and gold is applied.

Most Buddhist bronzes were made in this way. Small ones were cast in one piece, while large ones were made in several sections.

The technique of embossing consisted of hammering a sheet of copper or bronze on a wooden model. The method was used particularly for flat bronzes or small articles made of gold or silver. But pieces made in this way were much simpler and coarser; only the bronzes made by the *cire-perdue* process are of fine sculptural quality.

BUDDHIST FIGURES

Under the Six Dynasties Buddhist figures were confined to representations of the Buddha, bodhisattvas, minor divinities and Buddhist shrines. The Buddhas, usually Sakyamuni or the historical Buddha, and Maitreya or the Buddha of the future, appear in different postures (seated, standing or reclining), with a great variety of attributes and different gestures of the hands *(mudras)*. Guanyin [Kuan-yin] or Avalokitesvara, seen as an emanation of the Amitabha Buddha, was the most popular bodhisattva at this period.

Seated Buddhas

The Chinese wished to represent him in his characteristic forms; so Sakyamuni in the lotus position is a common figure. He is seated cross-legged on a throne in the gesture of meditation or *dhyana mudra* (the hands resting one on top of the other in his lap). This attitude of meditation, in which the figure is seated in the Indian fashion, is also known as *samadhi*.

The figure wears a monastic robe that falls in gentle folds. On the top of his head is the protruberance known as *ushnisha*, symbolizing his supernatural essence. His face expresses immense calm and confidence, which may be seen as a manifestation of the artist's mystical search.

The influence of the art of Gandhāra and sometimes even of the art of Mathura from northern India is very evident in fourth century pieces. One of many examples is the Sakyamuni Buddha in the Winthrop Collection in the Fogg Art Museum: the folds of the drapery, the face and the hair are purely Gandharan. This piece was certainly made in China but was clearly copied from, or inspired by, pieces from Gandhāra — except for the throne, where the iconography is very Chinese.

Sakyamuni with hands in *dhyana mudra* would appear to have been the most popular image during the early periods of Buddhism in China in the fourth and fifth centuries. Numerous statues, not all of which have been dated, are at present preserved in various museums and private collections.

Towards the end of the fifth century the fervour of the masses was addressed to another Buddha image. This was still the Sakyamuni Buddha but now making the gesture of the first sermon.

At the same period scenes from the Buddhist legends began to appear in low relief on the backs of statues. These bronze images executed by the Wei — who were the great promoters of Buddhist art in China — bear a strong resemblance to the exceptional sculptures in stone that they made at the same period at the Yungang and Longmen caves.

During the sixth century the Buddha's whole body became lost beneath a veil of linear folds; flames and flying *apsaras* appeared round the nimbuses. During the second half of the sixth century, which is marked by the decline of the Wei and the rise of the Northern Qi, a new wave of Indian art brought about a change. Style became more plastic and less linear under the influence of Gupta art. The figures usually represent Maitreya and Sakyamuni, seated either in the Indian or the European fashion or in *bhadrasana* (with both legs pendent).

Standing Buddhas

To judge from the known figures, it would seem that the standing Buddhas appeared later than the seated ones. In fact, there is no example of a standing Buddha of fourth-century date. The earliest standing figure, published by Munsterberg as Plate 19 of *Chinese Buddhist Bronzes,* is dated 443. It is a Maitreya Buddha standing on a lotus base in a style known as *Udayana*. He wears a monk's habit that falls in a series of concentric folds. His right hand is raised and open, palm forwards, in the gesture of absence of fear *(abhaya mudra)*. The fingers of the left hand are extended, and the arm is outstretched in *vara mudra,* which indicates the giving of the self by instruction. The hair is strongly reminiscent of Gandhāran art; the head is surmounted by the *ushnisha*.

However, the most characteristic figure of the fifth century is the standing Maitreya in the Metropolitan Museum of Art (see Pl. 113). This gilded bronze, measuring 1.385 metres in height and dated by an inscription to 477, is one of the most representative sculptures of the art of the Northern Wei of the second half of the fifth century.

Although it is executed in a *Udayana* style, strong Indian influence is apparent in the iconography; the handling of the drapery derives from Indian sources, while the head-dress and silhouette bear the stamp of Gandhāran art, which had been introduced into China via Central Asia. This Maitreya, like the figure already described, stands on a lotus base. He wears a monastic habit, its regular folds ending in pleats, which moulds the Buddha's body and reveals all the plastic beauty of the sculpture. Like nearly all the figures of the period, the left hand is in *vara mudra* and the right hand in *abhaya mudra*.

110

110
Standing Buddha on a lotus base with a flaming mandorla at his back.
Gilded bronze
Northern Wei dynasty, dated 520
Height 29 cm.
Asian Art Museum, San Francisco, Avery Brundage Collection

111
Buddha seated in the lotus position on a low base, hands resting in his lap in the gesture of meditation *(dhyana mudra).*
Gilded bronze
Later Zhao dynasty, dated 338
Asian Art Museum, San Francisco, Avery Brundage Collection

111

The sixth century is regarded as the high watermark of Buddhist sculpture in bronze; it saw the creation of the finest standing Buddhas. These include the Maitreya (dated 520) in the Detroit Institute of Fine Art and the Maitreya (536) in the University Museum of Philadelphia, as well as the two Maitreyas of the early sixth century in the Freer Gallery of Art. All these sixth-century figures are standing Buddhas on a lotus base with a great flaming mandorla at their backs. The heads have the *ushnisha,* and a smile of immense serenity plays over their features. The position of the hands is invariable: left hand in *vara mudra,* right hand in *abhaya mudra.* It is noteworthy that from beginning of the sixth century a new element was introduced, for the Chinese had freed themselves of foreign influences, and these statuettes are pure reflections of their own culture.

With the ending of the Wei dynasty in 534 figures of the Maitreya and Sakyamuni Buddhas became fewer, as they declined in favour, and the numbers of Amitabha (transcendant Buddha) and Vairocana (cosmic Buddha) figures increased.

Buddhas of the Past

Mahayana Buddhism teaches that Buddhas known as 'Buddhas of the past' preceded Sakyamuni, the historical Buddha. In China they number seven and correspond to the sun, the moon and the five planets.

According to the *Lotus Sutra,* one of the most popular and important Mahayana Buddhist texts in China, of all the Buddhas of the past, Prabhutaratna was held in highest repute. Prabhutaratna's fame is confirmed by the existence of large numbers of gilded bronze statuettes. He is most often represented seated beside Sakyamuni. Groups of this type are preserved in the Freer Gallery of Art, the Nezu Museum, the Musée Guimet, etc.

The group in the Musée Guimet, dated 518, is one of the subtlest depictions of the theme known. It is probably one of the best works executed under the Northern Wei, made in the finest style of the period. It is a masterpiece

112
Standing Maitreya Buddha on a lotus base.
Gilded bronze
Northern Wei dynasty, dated 536
Height 61.7 cm.
Museum, University of Pennsylvania, Philadelphia

112

that reflects the skill of the Chinese bronze-casters at the peak of Buddhist art. The two Buddhas sit in *lalitasana*, the pose of relaxation, (left leg bent, right leg pendent) conversing gravely, each with hands making the same gestures: right hand in *abhaya mudra*, left hand in *vara mudra*. The bodies are entirely lost beneath monastic robes outlined by sharp folds. The serene faces, radiant with the famous archaic smile, are set off by mandorlas decorated with a flickering flame motif. At the base of the throne are lions in high relief and monks in low relief on either side of the censer. Despite the fact that the figures are a quarter turned away from the spectator and for all the extreme delicacy of the execution, the frontal character of Wei sculpture is still perceptible.

Avalokitesvara or Guanyin [Kuan-yin]

After the Buddhas the bodhisattvas are the most sacred beings in the Buddhist pantheon. They play an important part in Mahayana Buddhism and are often known as *Mahasattva* (great beings). They are beings on the way to Enlightenment, destined to reach the state of Buddhahood soon.

Depicted either as attendant upon a Buddha or as a statuette on its own, a bodhisattva is always clothed like an Indian prince, adorned with a necklace and jewels, the head crowned. The most popular bodhisattva in China was Avalokitesvara, whom the Chinese called Guanyin. He was considered to be an emanation of the Amitabha Buddha, because he wears an image of jina Amitabha in his head-dress. Representations of this bodhisattva are legion, but the commonest form is Avalokitesvara Padmapani, i.e. Avalokitesvara holding a lotus stem (symbol of purity) in his right hand and a ritual vessel containing the elixir of life *(amrita)* in his left.

Munsterberg states that the earliest known statuette of Guanyin is the one preserved in the Fujii Museum, Kyoto. This fourth-century bronze depicts a standing bodhisattva, clothed like an Indian prince and adorned with jewels. His right hand is in *abhaya mudra;* in his left he

113
Standing Maitreya Buddha on a lotus base.
Gilded bronze
Northern Wei dynasty, dated 477
Height 138.5 cm.
113   Metropolitan Museum of Art (Kennedy fund 1926), New York

114

115

146

114
Stele representing the standing bodhisattva Avalokitesvara (Guan-yin).
Gilded bronze
Northern Wei dynasty, dated 513
Height 18.6 cm.
Eskenazi Ltd., London

115
Stele with standing bodhisattva Avalokitesvara.
Gilded bronze
Northern Wei dynasty
Height 26 cm.
Maurice Bérard Collection, Paris

holds the ritual *amrita* vessel. Like the earliest Buddha figures, this one is very close to the Gandhāran proto-types. The similarity of the moustache and the drapery are particularly striking.

However, Guanyin's popularity would appear to have developed in China during the second half of the fifth century. Many dated figures of that period have survived, but they are statuettes of Avalokitesvara in his guise of Padmapani.

Although very popular under the Six Dynasties, Guanyin or Avalokitesvara is only rarely depicted in a seated position.

After Guanyin the most popular bodhisattva is Maitreya. He is usually seated in meditation, cross-legged, his right ankle resting on his left knee, his head leaning on the tips of the fingers of his right hand. This position was highly regarded in Korea and Japan, and the finest examples come from these countries.

Few figures of this type have survived, and the example in the Musée Guimet described by Munsterberg as 'China, end of the sixth century' is now recognized as being Korean.

116
Stele with standing bodhisattva Avalokitesvara.
Gilded bronze
Northern Qi dynasty, dated 571
Height 32.5 cm.
Asian Art Museum, San Francisco, Avery Brundage Collection

116

117

117
Detail of Plate 118

118
Large Buddhist shrine. The central figure stands on a throne of reversed lotus and represents the Maitreya Buddha: his right hand in *abhaya mudra*, his left hand in *vara mudra*.
Gilded bronze
Northern Wei dynasty, dated 524
Height 59 cm.
Metropolitan Museum of Art (Rogers Fund 1938), New York

119
Arhat (luohan).
Gilded bronze
Six Dynasties, *c.* sixth century.
Fogg Art Museum (Grenville L. Winthrop Bequest), Cambridge, Mass.

Arhats or *Luohans*

These are not divine beings but saintly men following in the 'Master's' footsteps. These monks, usually in pairs, are placed on either side of the Buddha. They are bald-headed, clad in monastic robes, the palms of their hands pressed together in the gesture of adoration (*anjali mudra*).

Donors may be included in this category. They are images of those who hope to enter paradise and whose names are mentioned in the inscriptions.

Buddhist Trinities and Shrines

Trinities, consisting of a Buddha attended by two bodhisattvas, are common in the Buddhist art of China. They usually depict the Sakyamuni Buddha and the two bodhisattvas Avalokitesvara and Mahasthamaprapta.

Very rarely a bodhisattva occupies the central position of the trinity. In these exceptional cases the privilege falls to Avalokitesvara.

Although the earliest known examples of dated trinities belong to the Sui dynasty, it would seem — as the monumental stone trinity in cave XXII at Yungang may suggest — that the first trinities in bronze were made as early as the fifth century.

Buddhist shrines usually consist of a trinity surrounded by many other figures. All the figures are large, measuring between 60 and 80 centimetres in height.

What is probably the most interesting Buddhist shrine is preserved in the Metropolitan Museum of Art. The figure, measuring 59 centimetres in height, bears an inscription dating it to the fifth year of Zheng Guang

120
Sakyamuni Buddha seated in the lotus position.
Gilded bronze
China, end of the fourth-beginning of the fifth century
Height 16.1 cm.
Eskenazi Ltd., London

120

121

122

121
Stele with Sakyamuni and Prabhutaratna Buddhas seated in the lotus position.
Gilded bronze
Northern Wei dynasty, dated 472
Height 16 cm.
Asian Art Museum, San Francisco, Avery Brundage Collection

122
Sakyamuni and Prabhutaratna Buddhas seated side by side in conversation.
Gilded bronze
Northern Wei dynasty, dated 518
Musée Guimet, Paris

[Cheng-kuang] (one of the names of the reign of Emperor Xiao Mingdi [Hsiao Ming-ti], i.e. A.D. 524). The large central figure depicts the Maitreya Buddha standing on a throne of reversed lotuses; his right hand is in *abhaya mudra,* his left in *vara mudra.* The serene face, radiant with a gentle smile, is one of the great masterpieces of the Wei bronze-casters.

The head, surmounted by the *ushnisha,* is adorned on either side by two great ears that drop nearly to the shoulders. The figure is clothed in a full monastic habit falling in tiered folds and spreading in wing-like points. Behind it rises a mandorla of flames with halos at the level of the Buddhas's head, its outer edge adorned with flying *apsaras* holding musical instruments in their hands. On either side of Maitreya, but very much smaller, the two bodhisattvas Avalokitesvara and Mahasthamaprapta stand on a lotus base. At the Buddha's feet and on either side of the censer supported by the goddess of the earth are two standing *dvarapalas* and two Buddhist lions.

However, few groups of such importance and quality have come down to us, many having probably been destroyed during the persecutions of Buddhism in China.

With the introduction of Buddhism into China, there came a renewal of the art of bronze, which had begun to decline as early as the eighth century B.C. This art reached its peak under the Northern Wei (386–535) and the Northern Qi (550–77).

Under the Northern Wei the figurines are more delicate and better-proportioned. Faces are more Chinese, being long with high foreheads, and are suffused with the archaic smile characteristic of the period. Heads are sometimes adorned with a lotus-shaped crown. While a Greco-Indian influence is perceptible in the faces, the Greco-Buddhist art of Gandhāra is apparent in the broadly undulating head-dress.

The statuettes, clothed in garments with parallel folds or 'damp fold' drapery, stand on circular pedestals in the form of a lotus or of a throne supported by a socle. Mandorlas are often adorned with flying *apsaras* in place of flames. Figures related to the central statuette are sometimes engraved or sculpted in low relief on the mandorla. The votive images used for worship, of which an immense variety were made, reflect all the features of the Wei style as found in the caves of Longmen.

New trends made their appearance with the Northern Qi: plastic beauty was reinstated, images of greater humanity were sought and found, drapery became suppler and more transparent. This new state of mind in the artists stemmed from the undoubted influence of the Gupta art of India and from a reaction against the severe and linear style of the Wei. Thus sober garments, broken by slender fillets in relief or shallow incisions, fall in regular folds that terminate in scrolls. Despite the deeper incisions of the ornament, modelling has become softer and more rounded, so that figures can be given more flexible postures; the rounded modelling of the eyelids lends the faces a more human expression than those of the Wei; there is greater suppleness in hair and hands.

The last word shall be left to Daisy Lion-Goldschmidt, who wrote that the sculpture of the Northern Qi is a 'synthesis of the fullest plasticity and an intense expression of spirituality'.

124
Standing Buddha.
Gilded bronze
Sui dynasty, dated 589
Height 24 cm.
Asian Art Museum, San Francisco, Avery Brundage Collection

Growing religious fervour facilitated the spread of a wholly Chinese form of Buddhism, with new interpretations and doctrines.

BUDDHIST ART

Thanks to the attitude of the emperors who favoured Buddhism, thereby encouraging its naturalization and practice, numbers of Buddhist figures in stone and bronze greatly increased.

Sui statuary finally broke with the art of the Northern Wei and reacted against the Indian influence that was present in the art of the Northern Qi. Nevertheless, the bronze statuettes retained an impressive rigidity. The body, consisting of a solid block, is often ovoid. The head is also ovoid, with a fixed expression on the square face, and a heavy chin. There is also a pronounced tendency to load nimbuses, drapery, etc. with lavish ornament and a profusion of rosettes, lotus, garlands and jewels. Divinities are adorned with heavy crowns, long crossed necklaces and flowers. However, there are still rare exceptions, dating from the beginning of the dynasty, in the style of the Northern Qi. Innovations under the Sui may be summed up as follows: the overall appearance is cylindrical or ovoid; the edges of the eyelids form a supple line; there are three rolls of flesh round the neck; ornament is profuse.

125
Standing Amitabha Buddha; his right hand performing the *abhaya mudra*, his left the *vara mudra*.
Gilded bronze
Sui dynasty
Height 23.5 cm.
Formerly J. Daridan Collection; sold Sotheby's, London,
11 December 1979

125

126
Standing Avalokitesvara on a base in the shape of a reversed double lotus.
Gilded bronze
Early Sui dynasty, late sixth century
Metropolitan Museum of Art (Rogers Fund 1912), New York

Seated Buddhas

Figures of seated Buddhas are comparatively rare under the Sui. However, a very distinct change of style and silhouette is discernible in these statuettes. The Buddha is usually seated in the lotus position (cross-legged) wearing a monastic robe, the linear folds of which fall in a gradual cascade. An ovoid appearance is already apparent in figures of this type.

Standing Buddhas

Chinese Buddhist art under the Sui reached a peak of geometric form, creating statuettes that were slightly cylindrical and had slenderer silhouettes. The outer robe was left open to reveal the monastic habit falling gracefully from the torso in concentric curves. It should, however, be noted that the projections on the lower part of the drapery, which had been handled with such skill under the Six Dynasties, have entirely disappeared. Nor were the Sui satisfied with a change of style: they depicted figures performing new and unusual gestures such as the *vitarka mudra* or gesture of discussion.

The most typical figures of this period are undoubtedly those representing the Amitabha Buddha. Standing on a slightly cubic base, this Buddha has the cylindrical silhouette characteristic of the Sui. The monastic habit falls in tiers, forming curvilinear folds. The hair is curly and the face squarish. The three rolls of flesh around the neck, a characteristic feature, are clearly visible.

127
Buddhist trinity: standing Amitabha Buddha on a lotus base, his hands in *abhaya* and *vara mudras;* flanking him to left and right are the bodhisattvas Avalokitesvara and Mahasthamaprapta.
Gilded bronze
Sui dynasty, dated 597
Height 32.1 cm.
Freer Gallery of Art, Washington, D.C.

127

Buddhas of the Past

As under the Six Dynasties, the Buddha of the past the Sui revered most widely venerated and portrayed most often was Prabhutaratna. The best example of the type is the bronze stele in the Freer Gallery of Art, dated by inscription to the year A.D. 609 (Pl. 128). The scene is identical with that of the celebrated Wei shrine described in the previous chapter. Prabhutaratna is seated beside the Sakyamuni Buddha. Both Buddhas are seated in the lotus position, both make the gesture of *abhaya mudra* with the right hand and *vara mudra* with the left. Although the modelling has greater plasticity, and there is no linear effect in the drapery, the bodies are more block-like with a cylindrical tendency.

With the fall of the Sui dynasty, the theme of Sakyamuni seated beside Prabhutaratna seems to have lost favour with the Chinese. The scene appears to have been non-existent under the Tang, the Song and the Ming.

Avalokitesvara

These figures exist in great numbers, some inscribed and dated, but the poses are static despite a slight S-curve of the hips and always tend towards the cylindrical silhouette that was so greatly favoured at the time. The most majestic version is in the British Museum. This statuette of Avalokitesvara Cintamanicakra or Guanyin holding the *cintamani* (magic jewel) is dated 595. Standing on a base of a double lotus reversed, Guanyin is clothed in the rich attire of a prince and is adorned with many necklaces and jewels. He holds the *cintamani* in his right hand and the elixir of life in a bottle in his left. Behind his head is a nimbus of lotus-petals, decorated on the rim with dancing flames. Although the hip is thrust slightly forward, the silhouette of the statuette remains very cylindrical.

Buddhist Trinities

The central figure of the Buddhist trinities of the Sui dynasty is usually Amitabha, though his place is occasionally taken by other Buddhas such as Maitreya and Bhaishajyaguru. The finest Chinese bronze trinity steles date from this period. There was in fact a great vogue for Buddhist trinities under the Sui, for they were linked to

the cult of Amitabha and presumably shared the enormous popularity of this Buddha.

The quality and beauty of a stele in the Freer Gallery of Art dated 597 (Pl. 127) sum up all the delicacy and mastery that a Sui artist could bring to such a work. The central figure standing on a lotus base is the Amitabha Buddha with his hands in the *abhaya* and *vara mudras*. The outer robe falls open to reveal a monastic habit with curvilinear folds. The squarish face wears an expression of great serenity. Behind the head is a nimbus of flames or lotus petals. The central figure is flanked by Avalokitesvara on the left and the bodhisattva Mahasthamaprapta (the representation of the spiritual wisdom of the Amitabha Buddha) on the right.

Despite the simple, powerful forms of the three figures, the Sui cylindrical effect is distinctly visible.

Buddhist Shrines

Since Buddhist shrines are no more than trinities with many additional subsidiary figures, it is natural that the central figure should be that of the Amitabha Buddha. A perfect example is the very large group (77 cm in height) dated 593 in the Museum of Fine Arts, Boston. Amitabha is shown seated in the lotus position on a raised throne in the form of a double lotus reversed, his hands in the *abhaya* and *vara mudras*. His monastic habit falls in light folds, leaving the right shoulder uncovered. The Buddha's curly hair is surmounted by the *ushnisha*. A circular nimbus behind his head is adorned with flying *apsaras*. Standing on a lotus socle to his left, Avalokitesvara holds a pomegranate in his right hand. On a lotus socle to his right, head slightly inclined towards his two hands in *anjali mudra* (gesture of adoration), stands Mahasthamaprapta. The two bodhisattvas have nimbuses in the form of lotus petals. Behind them and on a smaller scale stand Ananda and Kasyapa. On the right, the former holds a begging bowl and a scroll in a sutra box; on the left, the latter holds an open sutra. Behind each *arhat* stands a

128
Buddhist stele with Sakyamuni and Prabhutaratna, Buddhas of the past, seated in the lotus position: right hands in *abhaya mudra*, left hands in *vara mudra*.
Sui dynasty, dated 609
Height 21.8 cm.
Freer Gallery of Art, Washington, D.C.

128

Pratyeka Buddha. In the foreground flanking an incense-burner supported by a goddess of the earth sit two lions with keepers behind them. Above the Buddha are the spreading leaves of the sacred tree adorned with lotus flowers and garlands of jewels; its tips are decorated with seven Buddhas of the past.

These Buddhist shrines with their dynamic composition, the head of the Buddha adorned with a circular nimbus, all standing out against a background of the leaves of the sacred tree, still retain the typical cylindrical appearance of Sui art.

Short-lived though it was, the Sui dynasty left a powerful mark on Buddhist art in China. These statuettes are exceedingly rare, but their quality and modelling are of such perfection as to bear comparison with the best Chinese sculpture in stone.

The Tang Dynasty

Chronology of the Tang [T'ang] Dynasty

Emperors

Gaozu [Kao-tsu]	618–26
Taizong [T'ai-tsung]	627–49
Gaozong [Kao-tsung]	650–83
Zhongzong [Chung-tsung]	684–709
Ruizong [Jui-tsung]	710–12
Xuanzong [Hsüan-tsung]	713–55
Suzong [Su-tsung]	756–62
Daizong [Tai-tsung]	763–79
Dezong [Te-tsung]	780–804
Shunzong [Shun-tsung]	805
Xianzong [Hsien-tsung]	806–20
Muzong [Mu-tsung]	821–4
Jingzong [Ching-tsung]	825–6
Wenzong [Wen-tsung]	827–40
Wuzong [Wu-tsung]	841–6
Xuanzong [Hsüan-tsung]	847–59
Yizong [Yi-tsung]	860–73
Xizong [Hsi-tsung]	874–88
Zhaozong [Chao-tsung]	889–903
Zhaoxuan [Chao-hsüan]	904–22

To non-specialists the pottery horses and other figures of the Tang dynasty have made it perhaps the best known of all periods of Chinese culture. Yet the art of this dynasty covers many fields, including ceramics, goldsmiths' work, sculpture and bronzes. This proliferation of the creative arts was one of the results of Chinese expansion, of contacts and commerce with foreign countries.

From the moment when the first Tang emperors ascended the throne, China had launched upon an era of conquests that culminated in the annexation of Eastern Turkestan as far as Pamir and the subjection of Tibet, Korea and Annam.

In the wake of her territorial expansion came many contacts and commercial exchanges—with Iran, India, Central Asia and perhaps even with Byzantium—resulting from the continual movement of caravans and the presence of foreigners (Persians, Turks, Tibetans, Japanese, Koreans, Greeks, Indians) in the great cultural centre that the capital Chang'an had become.

HIGH WATERMARK OF BUDDHISM

For nearly seven centuries Chinese thought was dominated by Buddhism. It was a Buddhism profoundly modified by Taoism and Confucianism. This development culminated in the flowering of a purely Chinese brand of the religion with new interpretations, new doctrines and many new sects representing all the Indian schools. The best known are the Jingtu [Ching-tu] or Pure Land sect founded by Shandao [Shan-tao] and devoted exclusively to the cult of Amitabha; and the Chan [Ch'an] sect (purely Chinese) which preached sudden illumination.

China during the period of the Sui and Tang dynasties (end of the sixth century to mid-ninth century) was both the most important centre of Buddhism and its second home. With its growing numbers of monasteries becoming centres of religious and secular culture, this period is regarded as the golden age of Buddhist sculpture. But it was the pilgrims and translators of the Tang period who made this one of the greatest eras in Buddhist history.

Pilgrims and Translators

The first pilgrimages and translations date from the mid-second century A.D. In c. 260 Zhou Shixing [Chou Shih-

hsing] set off for Khotan in search of the original text of the *Prajnāpāramitā (Perfection of Wisdom)*. At the same time Zhou Fahu [Chou Fa-hu], also known as Dharmaraksha, travelled to India in search of the Law. He returned to China in c. 265 with many texts and settled in Chang'an, where he spent the rest of his life translating them into Chinese.

During the fourth century several monks travelled westwards and returned with Buddhist texts. Although it has not been possible to attribute any book to them and most remain unknown, a few names—Sengqian [Seng-ch'ien], Sengzuan [Seng-tsuan] and Tanzhong [T'an-chung]—have survived.

The fifth century saw the beginning of the great pilgrimages in search of sacred texts and the visits to the holy places. The most celebrated pilgrim of the period was Faxian [Fa-hsien]. Faxian left Chang'an in 399 to search for the original texts of the *Vinaya* or disciplinary rules for a monastic community; he visited Kucha, Khotan, Kashgar, Kashmir, the region of Kaboul, the Indus valley and the towns of the Ganges. On his return in 412 he wrote an account of his journey known as *Memoirs of the Buddhist Kingdoms or Report by Faxian.*

In 518, accompanied by several Chinese monks, Song Yu [Sung-yü] left Luoyang for the high valley of the Indus. The mission was official: Song Yu was charged by the Empress Hu of the Northern Wei to obtain a collection of Buddhist texts. He returned in 522 bringing manuscripts of 170 works of the Great Vehicle.

Between 575 and 581 a group of eleven monks travelled to India. They returned to China with 260 works in Sanskrit, many of which were translated in the following years.

But the greatest of Chinese pilgrims and the most prolific of Chinese Buddhists belongs to the Tang period. The name of this eminent scholar was Xuanzang [Hsüan-tsang] (602–64). Having become deeply learned in Buddhism, he set off alone in 629 for the deserts of Central Asia in search of a manuscript of the great treatise on metaphysics known as *Yogācārya (Lands of the Masters of Yoga)* and determined to perfect and enlarge his knowledge of Buddhist philosophy. Having visited the holy places, traversed India from east to west and from north to south and studied under masters of the highest repute, he returned to China in 645 and was received by the Emperor Taizong. Not only did he take back 657 works containing mainly the sutras of the Great Vehicle and certain texts of the Small Vehicle but, most importantly, he had gained complete mastery of Sanskrit. He spent the last nineteen years of his life translating and overseeing the best known and most prolific teams of translators in the whole history of Chinese Buddhism. The Chinese owe to him seventy-five works or one quarter of all Chinese translations of Indian Buddhist texts.

At the end of the seventh century Yijing [I-ching] (635–713) sailed by the sea-route to India. On his return in 695 he was received at Luoyang by the Empress Zetain. Of the 400 works that he took back, he translated fifty-six, the most important being a sutra of the Great Vehicle and texts concerning discipline (Vinaya) from the Small Vehicle. His two major works were the *Nanhai jiqui neifa zhuan [Nan-hai chi-kuei nei-fa chuan]* (Report on Buddhism, Sent from the Southern Seas) and the *Datang Xiyu qiufa gaoseng zhuan [Ta-t'ang Hsi-yü ch'iu-fa kao-seng chuan]* (Narrative about the Eminent Monks Who Went in Search of the Law in the Countries of the West in the Time of the Great Tangs).

Numbers of pilgrimages began to dwindle in the eighth century. However, the monk Huichao [Hui-ch'ao] should not be overlooked; he wrote an account of his travels that contains information on ethnographic, political, linguistic and religious matters. And the monk Wukong [Wu-k'ung] made several translations.

Following the proscription of Buddhism in 845, and the fall of the Tang dynasty, pilgrimages ceased and the Buddhist religion declined.

SUPPRESSION AND DECLINE OF BUDDHISM

The revival of Chinese nationalism, stimulated by the reactions of the Confucian and Taoist literati, came to a head in 845 with a decree by the Emperor Wuzong proscribing Buddhism in China. It was followed by fierce persecution and suppression of this 'foreign' religion, culminating in the confiscation of the property of the sects and the destruction of over 4600 monasteries and 40,000 small establishments. Convents were closed, temples destroyed or transformed into public buildings, and bronze statues were sent to be melted down. The decree accused Buddhism of having caused the moral and economic enfeeblement of the southern dynasties, but in reality the causes were deeper. They included a reaction of the literati against the eunuchs, who were fervent Buddhists and controlled practically all the executive power, and the scandalous wealth of the monasteries in

land, men, money and metals. In fact, the monasteries possessed the major part of the Empire's precious metals in the form of statues, bells and ritual objects. Finally there were the difficulties within the state treasury and the shortage of copper for minting money. The rigorous suppression and consequent rapid decline of Buddhism in China occurred in stages: first, the clergy were purged in order to eliminate spurious and untutored monks; then the bonzes' private possessions were confiscated; Buddhist ceremonies that had been introduced into the official services were suppressed; finally a general inventory of monastic possessions was made and all the possessions were confiscated.

BUDDHIST SCULPTURE

Exposed to a new wave of Indian influence, the Buddhist sculpture of the Tang period was the culmination of the development of the art of the Qi and the Sui. Despite the attempt to express a static reality (the movement and action of the human body), the small bronze statues continued to mirror the sculpture of the Buddhist caves. By the beginning of the eighth century, the variety of attitudes, freedom of movement and the abandonment of frontality in the silhouettes brought a plastic quality to the volumes.

Seated Buddhas

There are comparatively few gilded bronze figures of the seated Buddha from the Tang period. But existing examples include figures of the Sakyamuni, Amitabha, Vairocana and Maitreya Buddhas.

In Munsterberg's opinion the finest seated Buddha of the seventh century is the figure in the Metropolitan Museum of Art. It depicts the Sakyamuni Buddha seated in the lotus position, his hands in *dharmacakra mudra*

129
Very rare funerary mask of a prince (?)
Bronze with a green patina
Liao dynasty, tenth century
Height 21 cm.
Gisèle Cröes Collection, Brussels

130
Back-view of the Vairocana Buddha (see Plate 131). Scenes of the Buddhist hell with demons torturing the souls of the dead in low relief on the back of the figure.
Early Tang dynasty, seventh century
Height 14.1 cm
Musée Guimet, Paris

131
Vairocana Buddha; his right hand performing the *abhaya mudra,* his left the *vara mudra.*
Gilded bronze
Early Tang dynasty, seventh century
Height 14.1 cm.
Musée Guimet, Paris

130

131

(gesture of turning the Wheel of the Law or preaching the Buddhist gospel).

From the eighth century, figures of the seated Buddha began to differ greatly in style and posture, i.e. in the folds of the habit and in the positions of legs and hands. *Mudras* that had not been used in earlier periods came into vogue under the Tang. During the eighth century also, the Buddha began to be seated in the half lotus position (one leg over the other, with only one foot visible), whereas during the seventh century, under the Six Dynasties and the Sui, he had been seated in the lotus position (cross-legged, with both feet resting on the upper surface of the thighs).

The figure of the Buddha seated in the European position became equally popular. Here the Buddha is seated on a throne, with both legs pendent, one hand on his knee, the other in *abhaya mudra* (Pl. 132).

In Buddhist art of the Tang period specific gestures are characteristic of certain Buddhas. For example, a Buddha making the gesture of *dhyana mudra* is always Amitabha (Buddha of the Western Paradise) and a Buddha in *vara mudra* (symbol of the world of diamond and the world of darkness, symbols of male and female) is always the cosmic Buddha Vairocana.

Standing Buddhas

In its images of the standing Buddha, Tang art returned to the fountain-head of the 'foreign' religion, drawing its inspiration from the more plastic and sensual Indian style. Forms became more naturalistic; features of bodies and faces more strongly marked. Openwork nimbuses were adorned with flying *apsaras*. Plinths were decorated with an incense-burner and Buddhist lions.

The standing Amitabha was undoubtedly the commonest and most popular image under the Tang. Other Buddha figures that were the object of great religious fervour at that period include Bhaishajyaguru (Buddha of medicine), known in China as Yaoshi, characterized by a golden *haritaki* fruit or a medicinal jar held in his hands, and Vairocana, whose images are the most interesting of the standing figures from the point of view of iconography. The Musée Guimet possesses a very fine example

132

132
Maitreya Buddha seated in the European posture, right hand resting on his knee, left hand raised in *abhaya mudra*.
Gilded bronze
Tang dynasty, seventh or beginning of eighth century
Height 6.5 cm.
Eskenazi Ltd., London

133
The seven Buddhas of the past.
Gilded bronze
Tang dynasty, seventh century
Height 22.5 cm.
Asian Art Museum, San Francisco, Avery Brundage Collection

(Pls. 130−1) with particularly attractive ornament. Scenes of the Buddhist hell, where demons torture the souls of the dead, are sculpted in middle relief on the back of the figure, while on the front are scenes depicting the joys of paradise. The solar disc on the Buddha's shoulders contains a bird and a crescent moon, symbolizing the sun and moon respectively.

The most important point about the standing Buddhas of the Tang dynasty − and, indeed, the whole of Buddhist art of the period − is that it reflects a revival of Indian influence combined with an advance in the art of bronze.

Buddhas of the Past

As has been noted in the chapter on the Six Dynasties, Mahayana Buddhism teaches that in addition to Sakyamuni (the historical Buddha) and Maitreya (the Buddha of the future) there are 'Buddhas of the past' who preceded Sakyamuni. They are usually said to number seven, the figure corresponding to the sun, the moon and the five planets, but according to other beliefs there are five Buddhas related to the five directions (four cardinal points plus the centre), and three other Buddhas who also preceded Sakyamuni. These different versions account for the fact that Buddhas of the past may occur in groups of three, five or seven.

The Tang images of the Buddhas of the past are considerably less interesting than those of the Wei and the Sui. There are no groups depicting Prabhutaratna seated beside Sakyamuni, for both had ceased to be popular with the people at large.

134
Standing Avalokitesvara, in a slightly S-curved posture, on a lotus base.
Gilded bronze
Tang dynasty, end of the seventh-beginning of the eighth century
Height 28 cm.
Asian Art Museum, San Francisco, Avery Brundage Collection

135
Bodhisattva Avalokitesvara, or Guanyin, seated in *lalitasana*.
Gilded bronze
Tang dynasty, beginning of the eighth century
Height 10.5 cm.
Eskenazi Ltd., London

134

135

Buddhas of the past under the Tang are usually, if not invariably, elements in the mandorla or nimbus of the statuettes.

Avalokitesvara (Guanyin)

According to Buddhist legend, this bodhisattva of mercy and compassion, who has delayed his entry into Nirvana in order to save souls, is an 'ocean of pity' who assists the needy in their hour of peril and a protector on whom the people may call at moments of distress. He is thus the last resort.

Although J. Hackin has counted over 108 forms of this bodhisattva, he is readily identifiable by the presence in his head-dress of a tiny image of the Amitabha Buddha, of whom he is the emanation.

Under the Tang, figures of this bodhisattva were more numerous than those of any other divinity or Buddha. This popularity is probably explained by the importance attached in China to the *Lotus Sutra*. The style aims solely to recreate the plastic beauty of the body and draws its inspiration directly from Indian art. Although there are many variants, the main Tang images of Avalokitesvara are a standing statuette with eleven heads, indicating that the bodhisattva can look in all directions at once (this means that he is the refuge for all since he can see all those who implore his aid. The eleven heads, which occur frequently in Indian art, are the mark of a supernatural being); a standing statuette with four arms: two holding the lotus and the bottle of elixir, the other two in *abhaya* and *vara mudras* respectively; a standing statuette, the body or torso forming a curved line that creates a 'dancing' triple-bend movement known as *tribhanga* (this is the most typical of the Tang bodhisattva images); seated statuettes; these were as popular under the Tang as standing figures. There were two possible positions: one with legs crossed in the lotus position; the other with one leg in the half-lotus position, the other pendent in the position of relaxation or *lalitasana*.

Bodhisattvas other than Guanyin

After Avalokitesvara the main bodhisattvas venerated under the Tang were Maitreya, always shown seated in meditation (see p. 172), Manjusri, bodhisattva of wisdom, known as Wenshu [Wen-shu] in Chinese. After

136

136
Manjusri, boddhisattva of Wisdom, riding a lion.
Gilded bronze
Tang dynasty, seventh-tenth century
Musée Guimet, Paris

137
Dancing bodhisattva, the body in the S-curved *tribhanga* position.
Gilded bronze
Tang dynasty, eighth century
Height 12.7 cm.
Musée Guimet, Paris

138

138
Seated figure of the Kshitigarbha Buddha with the *cintamani* in his right hand.
Gilded bronze
Tang dynasty, sixth-seventh century
Courtesy of the Museum of Fine Arts, Boston

recognized by his shaven head, monastic habit and his attributes, the *khakkhara* and the *cintamani* (magic jewel); Samantabhadra or Puxian [P'u-hsien], venerated as combining the highest intelligence with good actions; usually shown riding an elephant. This bodhisattva appears to have been particularly popular under the Ming; Vajrasattva or 'essence of the thunderbolt'. This bodhisattva holds the *vajra* (symbol of Buddhist knowledge that destroys evil passions) in one hand and the *ghanta* (a small bell) in the other; Akasagarbha, a bodhisattva who is the 'essence of emptiness'. This is the most esoteric form of Tang Buddhism. His attributes are a pearl or jewel, a sword or a three-pointed *vajra*; unidentified bodhisattvas in various positions: standing, dancing, seated or kneeling on one knee. The Musée Guimet possesses a fine example of a dancing bodhisattva. The S-curve of the body, the position of legs, feet and arms are unquestionably Indian in inspiration.

Minor Divinities

After the Buddhas and bodhisattvas, the pantheon of Mahayana Buddhism includes divinities regarded as demi-gods or holy figures. They are Vajrapani or protector-god, wielder of the thunderbolt. The most important of the minor divinities, his role is to guard the approach to temples. He is a very muscular individual with a ferocious air who is expected to drive away demons; *dvarapalas* or defenders of Buddhism, always in pairs. They often carry the *vajra* (symbol of the destructive force that is itself indestructible) as an emblem. The *dvarapalas* use this weapon to combat the enemies of Buddhism (human vices and passions and demoniac forces). These guardians are always depicted in dramatic postures, their S-curved bodies muscular and uncouth, with hideously ferocious expressions. They are sometimes clad in armour like the warriors of the period; *lokapalas* are always four in number as their duty is to guard the four corners of Buddhist shrines; (as the iconography of all these guardians is very similar, it is extremely difficult, and sometimes impossible, to differentiate between them when they are divorced from their original context); *apsaras*. The *apsaras* or Buddhist angels are the most attractive of the minor divinities. They are simply attendant figures or decorative elements on Buddhist steles and thus suggest an image of paradise. *Apsaras* are usually depicted with agreeable feminine features and play musical instruments.

Avalokitesvara he is the most important personage in the *Lotus Sutra*. He is shown riding a lion or holding a book or a sword in his hand; Kshitigarbha, known as Dizang [Ti-tsang] in Chinese. He appears often in the Dunhuang paintings. Although very rare in gilded bronze, he can be

139

139
Standing *lokapala*.
Gilded bronze
Tang dynasty, seventh-tenth century
Musée Guimet, Paris

140
Figure of an *arhat* holding the sacred vessel in his hands.
Gilded bronze
Tang dynasty, seventh century
Height 13.6 cm.
Freer Gallery of Art, Washington, D.C.

140

179

Arhats and Donors

Called *luohan* by the Chinese, *arhats* are not divine but occupy a position in the Buddhist pantheon that somewhat resembles that of the twelve apostles of Christ in Christianity; however, they number sixteen, eighteen and sometimes 500. These monks with shaven heads, clothed in monastic habits, very often with hands in *anjali mudra* (gesture of adoration) or carrying a sacred vessel, appear in pairs on either side of a Buddha or a trinity. They are usually figures of Ananda and Kasyapa, Sakyamuni's two favourite disciples and the first Buddhist patriarchs.

There are no characteristic portrayals of donors; they are usually figures of men or women in the gesture of adoration or of giving.

Buddhist Trinities and Shrines

As has been seen in the chapters on the Six Dynasties and the Sui, the Buddhist pantheon would not be complete without trinity groups and shrines.

The Musée Guimet possesses a fine trinity in gilded bronze. The three figures standing on lotus socles are those of Avalokitesvara flanked by two bodhisattvas. Guanyin (Avalokitesvara) is twice the size of his attendants and is easily recognizable from the little statuette of the Amitabha Buddha that adorns his head-dress and from the lotus-flower and elixir bottle he holds in his hands.

Few Buddhist shrines in gilded bronze of the Tang period have survived the centuries and the violent persecutions of Buddhism in China. Those that have

141
Buddhist trinity in gilded bronze; in the centre the bodhisattva Avalokitesvara holding a lotus flower.
Tang dynasty, seventh-eighth century
Height 22 cm.
Eskenazi Ltd., London

142
Buddhist shrine with the Maitreya (?) Buddha seated in the European posture, his right hand in *abhaya mudra*.
Gilded bronze
Tang dynasty, end of the seventh century
Height 32 cm.
Asian Art Museum, San Francisco, Avery Brundage Collection

141

143

143
Seated Ganesha, the elephant-headed god.
Gilded bronze
Tang dynasty, seventh-tenth century
Height 10.2 cm.
Cleveland Museum of Art (gift of Mr and Mrs Leroy B. Davenport),
Cleveland, Ohio

come down to us are small (25–30 cm), but they are not comparable in quality and elegance with those of the Northern Wei or even the Sui.

Sculpturally the most beautiful and iconographically the rarest group is undoubtedly the shrine in the Asian Art Museum, San Francisco. The central figure, seated in the lotus position on a raised throne of a double reversed lotus, is the Bhaishajyaguru Buddha (Buddha of medicine). His right hand is in *abhaya mudra* (absence of fear), the left in *bhumisparsa mudra* (taking the earth to witness). The openwork nimbus at his back is decorated with the seven Buddhas of the past. The bodhisattvas Suryaprabhasa (bodhisattva of the sun, holding the solar disc in his left hand) and Candraprabha (bodhisattva of the moon, holding the crescent moon in his left hand) stand on either side of the Buddha. In front of the Bhaishajyaguru Buddha and flanking the incense-burner are two monks or *arhats*, one making the gesture of adoration, the other that of sacrifice.

Miscellaneous Buddhist Bronzes

In addition to these standard figures of Chinese Buddhist art, there are certain figures that do not fit into this classification and must be classed as 'miscellaneous'. They are Budai [Pu-t'ai] or laughing Buddha, represented as a corpulent figure, fond of good living, his face wreathed in smiles. He is often surrounded by children, symbolizing prosperity and contentment; Laozi [Lao-tzu], the patron saint of Taoism presented in a style strongly influenced by Buddhist art: he stands on a lotus socle, a nimbus behind his head and a lotus in his hand; Ganesha, the elephant-headed god, much venerated in India; demons or hideous personages, directly inspired by Indian art; and lions. The latter occur frequently in Buddhist art and were of Indian origin but were influenced by the Middle East. The lion was the emblem of the clan from which Sakyamuni was descended. Lions usually appear before the throne on bronze steles. The large lions in gilded bronze that have been found on their own may perhaps be relics of enormous bronze groups destroyed in former times.

The Song, Yuan and Ming Dynasties

Chronology of the Song, Yuan and Ming Dynasties

Song [Sung] Dynasty	**960–1279**	
Northern Song	960–1127	Capital: Kaifeng
Southern Song	1127–1279	Capital: Hangzhou
Yuan [Yüan] (Mongol) Dynasty	1280–1368	Capital: Taidu [T'ai-tu], Peking
Ming Dynasty	1368–1644	Capitals: Nanking to 1420; Peking from 1420

As at the end of the Zhou dynasty, the fall of the Tang dynasty was followed by a decline in the art of bronze in China. However, great encyclopaedias were compiled, inventories were made and ancient ritual vessels published by eminent Song archaeologists in an effort to re-establish and revive the ancient rites and traditions, with the result that there was a renewal of the art of bronze and the casting of vessels in the classical style.

RITUAL VESSELS

Under these three dynasties ritual vessels were cast in the classical style, i.e. in the same spirit as the ancient vessels of the Shang and Zhou periods. Watson dismisses these pieces as archaizing and therefore of little importance, uninteresting and artistically sterile. In reality they are vessels inspired by ancient forms and ornament but bearing the stamp of the later periods. Thus all the ancient forms—*gu, jue, gui, hu* and *ding*—reappear, as does classical ornament, including the *taotie* mask, *leiwen* scrolls and *kui* dragons. Yet the meanders, cicada-wing motifs, dragons and *taotie* differ slightly from those of the early periods. The iconographic repertoire has become modified and stylized and so has moved away from the deep religious feeling it expressed in the Shang and Zhou periods. Ornament is sometimes heightened with inlays of gold and silver which give the vessel a certain elegance.

These vessels, which are of poor quality compared with their precursors, are indifferently cast, and the alloys are dissimilar. Metallographical examination would suggest that these bronzes, especially those of the Ming period, have a high zinc content. In some cases, where the proportion of zinc exceeds 20 per cent, the material ceases to be bronze and is brass instead.

A second category of ritual vessels comprises pieces of novel and baroque form or morphologically derived from the ancient bronzes. They may bear Arabic inscriptions, imperial marks (mainly Xuande [Hsüan-te]) or the signature of the bronze-caster. Such pieces are often undecorated and have a brownish-red patina that is sometimes flecked with gold.

But the problem of dating the vessels of either category and of attributing them to the Song, Yuan or Ming period is an extremely delicate one and may even be insoluble. Dated pieces and those that carry a reign mark are often later than the date indicated. For example, vessels bearing the Xuande mark continued to be made up to the eigh-

144

144
Fang lei decorated in an archaizing style with *taotie* masks and *kui* dragons.
Bronze inlaid with gold and silver
Song dynasty, tenth-twelfth century
Height 31 cm.
Private collection

145

145
He kettle in an archaizing form, decorated with bands of stylized *kui* dragons.
Bronze inlaid with gold and silver
End of the Song dynasty, twelfth century
Private collection

146
Hu with serpentine geometric motifs.
Bronze inlaid with solid gold
Song dynasty, eleventh-twelfth century
Height 42.5 cm.
Victoria and Albert Museum, London

146

147
Hu decorated in an archaizing style with *kui* elephants
and birds.
Bronze with a black patina
End of the Ming dynasty, seventeenth century
Height 62 cm.
Private collection

148
Ritual vessel combining the body of a phoenix with
the head of a dragon and decorated with *taotie* masks
and spirals.
Bronze inlaid with gold and silver
Song dynasty, twelfth-thirteenth century
Height 32 cm.
Victoria and Albert Museum, London

147

148

teenth century. The quality of the work alone may perhaps facilitate the dating of a piece to the Song rather than the Ming period. Yet it is not unusual even in the great museums to observe that there have been second thoughts over the dating of a piece which may for years have been regarded as Ming, is then dated to the Song, only to reappear a little later with the Ming label in position once more.

Scholars do not care to admit that practically nothing is known about certain Song, Yuan and Ming vessels. Yet the various compilations published under the Song and the Ming reveal that ancient vessels were collected by imperial order from all parts of China. This was the expression of an infatuation with the past and was reflected in the art of bronze by a return to origins which was presented as an innovation. It took concrete shape at the end of the Northern Song with the creation of vessels, ancient in shape and ornament, but combined with certain contemporary characteristics, thus producing somewhat bizarre replicas.

The most interesting of the archaizing vessels are decorated with gold and silver inlay. But when closely examined they are soon seen to be the product of Shang or Zhou iconographic motifs combined with the inlaying techniques of the Warring States and the Han. These somewhat baroque pieces are probably late, i.e. after the thirteenth century.

BUDDHIST BRONZES

Because of the decline of Buddhism in China following the persecutions under the late Tang, few Buddhist bronzes were made under the Song, and they are unusual, most commonly representing the bodhisattva Guanyin. Pieces of this type of Ming date are comparatively rare and of poor quality, though the influence of Tibetan art prompted a minor revival of gilded bronze.

Seated Buddhas

Examples that can be dated to the Song are extremely rare. Ming pieces are common but bear traces of a decline in Buddhist sculpture. The figures are mannered, with a hard, dry appearance that is characteristic of Ming sculpture.

149

149
Avalokitesvara seated in *maharajalilasana* or the position of 'royal relaxation'.
Gilded bronze
Song dynasty, tenth-twelfth century
Height 18.5 cm.
Ashmolean Museum, Oxford

150
Standing Guanyin or Avalokitesvara, strongly influenced by Indian Gupta art.
Gilded bronze
Song dynasty, twelfth century
Height 51.5 cm.
Metropolitan Museum of Art, New York

Standing Buddhas

These figures were rare or non-existent under the Song. Under the Ming they were banal, executed in a dry, severe style, lacking artistic inspiration and reflecting no spiritual feeling.

Guanyin or Avalokitesvara

These are the commonest figures under the Song. The finest pieces were made of wood. The bronze statues are merely bad copies of the marvellous figures sculpted in wood. However, the drapery is more refined than in earlier periods. Standing figures sometimes bear traces of Gupta influence, with the right hand in *vitarka mudra* and the left hand holding a lotus branch. Except for a few iconographic variations, these pieces resemble Indian statuettes. The points of resemblance are in the form of the clinging garments, the bare breast, the adornments — necklaces, bracelets — at arms and wrists, and in the hair, the earrings and the conical head-dress with its figure of the jina Amitabha. These statuettes are closer to Indian than to Chinese sculpture. The seated figures are represented in the position of royal relaxation or *maharajalilasana*, i.e. right foot on the plinth, right arm resting on the right knee, left leg pendent, the left foot resting on a lotus. This posture is very common in the sculpture in wood of the Song dynasty and also appears under the Ming. The elegant but languid air expresses repose, hence its designation: relaxation.

With the Ming came a minor revival of the art of Buddhist bronzes in the figures of bodhisattvas. The elaboration of forms, the richness of ornament and wealth of detail give the statuettes a meticulous and confused appearance. The seated images are strongly influenced by the art of Tantric Buddhism. This is expressed in the very stylized treatment and ornament which is Tibetan rather than Chinese. These figures sometimes carry a Chinese inscription. The finest typically Tibetan examples were made at the beginning of the Ming under Emperor Yongle [Yung-lo] (1403–24). These figures in gilded bronze, usually portraying bodhisattvas, are seated in the lotus position on bases of a double-petalled inverted lotus. The delicacy of the execution, the richness of the decoration and ornament reflect the new religious fervour of the Chinese at the beginning of the Ming dynasty, resulting from the establishment of monasteries of lamas in certain large towns in China and in Peking.

150

151
Small statue of the bodhisattva Amitayus seated in *padma-asana* on a lotus base, the double row of petals chased and engraved. Ming dynasty, beginning of the fifteenth century, probably reign of Yongle
Height 42 cm.
Private collection (sold C. Boisgirard and A. de Heeckeren, Paris, 22 April 1980)

152
Winged lion inlaid with gold and silver
Song dynasty, eleventh-twelfth century (?)
Height 21.5 cm.
British Museum, London

151

193

Other statuettes of Avalokitesvara made under the Ming were influenced by Indian art. Some Ming Guanyin figures, for example, are shown with eighteen arms or holding numerous attributes symbolizing the supernatural power of the bodhisattva. More rarely the figures have a thousand arms or eleven heads or portray other esoteric subjects.

Bodhisattvas other than Guanyin and Minor Divinities

After Avalokitesvara the most popular bodhisattva under the Song was Kshitigarbha. Although only very rarely represented in gilded bronze, he is easily recognizable from his shaven head, monastic habit and from the attributes, the *khakkhara* and the *cintamani,* which he holds in his hand.

After Guanyin the two bodhisattvas who were most highly venerated under the Ming were Manjusri, the bodhisattva of wisdom, who holds a book or a sword in his hand and is often shown seated on a lion, and Samantabhadra or Puxian [P'u-hsien], who may be recognized by the recumbent elephant on which he is seated.

As in earlier periods, so under the Song and the Ming, a certain number of Buddhist gilded bronzes representing *dvarapalas, arhats,* Budai, as well as various Taoist and Confucianist personages — such as Laozi and Guandi [Kuan-ti], god of letters — were made.

As things stand now, little is known of the art of bronze in the Song, Yuan and Ming periods, for — at least as regards the ritual vessels — these pieces were long dismissed as indifferent copies of ancient vessels. Moreover they are comparatively recent, and it was natural that the mainstream of study and research should have been directed towards the early periods in an attempt to understand and identify the rites of Shang and Zhou society. Study of Song, Yuan and Ming bronzes has suffered as a consequence. That at these periods the art of bronze was a relatively minor art cannot be forgotten, for the finest flowering of Song, Yuan and Ming art was undeniably in their pottery and porcelain.

Appendices

SHAPES

Before discussing the technique of bronze casting, it will be as well to look at shapes, for the expression 'ancient Chinese bronzes', which is applied only to the Shang and Zhou [Chou] dynasties, covers a great variety of ritual vessels, bells, weapons, ornaments for chariots, agricultural implements, etc. However, in view of the vastness of the subject, discussion will be limited to the shapes of vessels and of musical instruments.

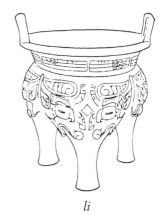

li

li

Nomenclature

The names of the ritual vessels are those which are mentioned in the inscription on a given vessel, occur in ancient texts describing the rites, were given by the Song [Sung] archaeologists, who were the earliest students of the ancient vessels, or are arbitrary names taken from the *Shuowen [Shuo-wen]* and the *Erya [Erh-ya]* (dictionaries compiled under the Han dynasty that describe the different shapes of vessels and their ritual functions).

Shapes

The shapes of ritual vessels have been studied by many scholars from the twelfth century onwards. By collating the results, Mizuno has been able to divide them into five categories: vessels for serving and holding food, cooking vessels, wine vessels, water vessels and musical instruments. In the following classification a distinction will be made between vessels for warming wine and those from which it was served and drunk.

Cooking Vessels

Li

Cauldron without cover, with a wide mouth, for cooking cereals and meat. It has three hollow legs which join to form the three-lobed body.

It was used continuously from the mid-Shang period until the beginning of the Period of the Spring and Autumn Annals.

Ding [Ting]

The *ding*, in all probability the earliest type of vessel cast by the Shang, is undoubtedly the most important of the Chinese bronzes. This tripod, used to cook food and cereals, has a hemispherical body, two vertical handles and three solid feet, more or less cylindrical.

The earliest shape, found at Zhengzhou [Cheng-chou] and dating from the end of the fifteenth century or the beginning of the fourteenth century B.C., has a very deep body and asymmetrical splayed feet, with rounded ends.

Under the Zhou the feet became thicker and the vessel shallower and more squat.

With the Periods of the Spring and Autumn Annals and the Warring States, the *ding* acquired a cover, which was sometimes surmounted by little animals in full relief. The handles were no longer on the rim of the vessel but were attached to the body itself; the feet were curved to resemble animals' paws. According to the *Erya* 'a very large *ding* is called *nai*; a *ding* with a narrow mouth is a *zai [tsai]*; a *ding* with handles on the outside of the body is a *yi*.' The two characters *nai* and *zai* occur in inscrip-

ding   *ding*

tions on bronzes of the Zhou period but do not correspond to the definitions given above.

The category *ding* includes several variants: knife-blade *ding*, square *ding (fang ding)* and *liding*.

Knife-blade *Ding*

A vessel similar to the *ding* but with three (or four) legs shaped like knife-blades. Mizuno believes that this type was confined to the late Shang.

Fang Ding [Fang-ting]

As its name suggests, this rather rare form is derived from the *ding* and has a square body standing on four legs. Until 1974, the *fang ding* was believed to have made its appearance in about the twelfth century B.C. or possibly at the beginning of the eleventh century B.C., but the excavations of 1974 at Zhengzhou, during which two *fang ding* were discovered, re-opened the question of date. Now it can be stated that the square *ding* was used from the end of the fifteenth or the very beginning of the fourteenth century B.C.

The Si mu wu *ding* excavated at Wu guancun [Wu Kuan-ts'un], near Anyang, measuring 1.33 metres and weighing 875 kilograms, is probably the largest so far known.

A few rare examples of this type of *ding*, such as the one exhibited by Eskenazi in London in June 1975, are what might be called 'knife-blade *fang ding*', for instead of being cylindrical the legs are shaped like knife-blades.

Liding [Li-ting]

This designation, re-used by Karlgren, is nevertheless mentioned in an inscription. The translation of the text mentioning a *liding* is controversial because the translations '*li* plus *ding*' or '*liding*' are both possible; however, the term will be retained here to describe vessels of intermediate shape between the *li* and the *ding*. This hybrid type is not quite hemispherical, has three shallow but distinct depressions above the legs and three hollow legs that are often cylindrical.

Xian [Hsien] or Yan [Yen]

This three-legged vessel was used to steam food and consists of two parts: the lower part, similar to the tripod *li,* contained the water; the upper part, in which the rice

xian

was placed, was called *zeng [tseng]*. Inside this part was a grid, either movable or fixed, called *bai [pai]*.

A four-legged form of *xian* is a considerable rarity; it has a rectangular body standing on four legs.

It continued to be made until the Han period, though with modifications to the shape of the vessel, and in this late period the lower element was shaped like a legless bowl.

Vessels for Serving and Holding Food

Gui [Kuei]

This vessel is called *duan [tuan]* in the inscriptions on bronzes and consists of a round cup standing on a ring foot. Examples occur with and without handles. Where handles occur they are semi-circular and are attached to the body of the vessel; there are two, or, very rarely, four.

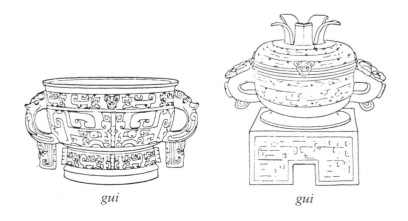

gui   *gui*

Mizuno believes that the type without handles made its appearance at the beginning of the Shang phase of Anyang (thirteenth-twelfth century B.C.), while the *gui* with handles are later (*c.* twelfth century B.C.).

Under the Zhou (end of the eleventh century B.C.), the ring foot stood either on three legs or on a solid cubic stand that was often larger than the vessel itself.

The *dui [tui], fu, xu [hsü], hui* and *yu [yü]* are often included in this group.

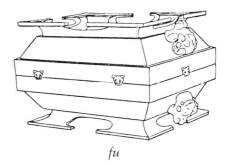

Fu   *fu*

An oblong rectangular dish with sloping sides, standing on a ring foot. It has a cover of the same shape and size as the dish.

The *fu* made its appearance under the Zhou towards the end of the ninth century or the beginning of the eighth century B.C. and remained in use until the end of the fifth century B.C.

Xu [Hsü]

This vessel somewhat resembles the *fu*, but the corners are rounded and the shape of the cover is very different from that of the body of the vessel. The body alone has two handles.

This type of vessel made its appearance in the middle of the period of the Western Zhou, towards the end of the ninth century B.C., and disappeared well before the Period of the Spring and Autumn Annals.

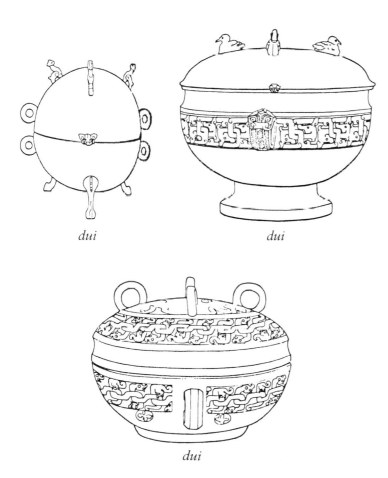

dui   *dui*

dui

Dui [Tui]

The term was first used by the Song to describe a category of round food vessels with cover. Rong Geng [Jung Keng] believes the *dui* to have been a spherical *gui* with a cover. But the *Erya* has the best definition: 'The *dui* is a vessel like the *fu* and the *gui*, the difference lies in the fact that it is entirely spherical.'

The *dui* probably made its appearance at the end of the sixth century B.C. and continued in use until the mid-fourth century B.C.

hui

Hui

The *hui* is a *gui* with a cover.

Yu [Yü]

This is a *gui* with L-shaped handles attached below the mouth of the vessel. The name is mentioned in inscriptions.

yu

Chinese archaeologists consider that the *yu* made its appearance at the end of the Anyang phase (twelfth-eleventh century B.C.) and disappeared at the beginning of the Zhou.

dou

Dou [Tou]

A semicircular bowl on a tall stem with splayed foot, usually with a cover that may be inverted to form a separate bowl.

The *dou* had occurred in pottery from the time of the Longshan [Lung-shan] culture and in the form of white pottery at Anyang; Watson believes that it made its appearance in about the ninth century B.C. It was in vogue during the Periods of the Spring and Autumn Annals and the Warring States and was abandoned at the beginning of the Han dynasty.

Vessels for Warming Wine

Jue [Chüeh]

This three-legged vessel with a handle on one side has a wide lip for pouring, a pointed extension and two vertical

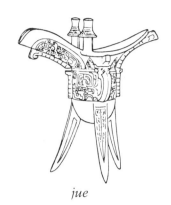

jue

columns surmounted by a sort of 'cap' or 'knob'. The body is a pointed oval in cross section. The *jue* was used over a low fire to evaporate what was left of the wine after it had been used for libations. There are several variants: some have a single cap, and the rarest *jue* have four legs.

The earliest examples, dating from the Zhengzhou phase (end of the fifteenth-beginning of the fourteenth century B.C.) have a flat base, a flattened body and thin walls. Some of the *jue* from about the twelfth century B.C. had covers; the specimen B60 B1049 in the Avery Brundage Collection in San Francisco is one of these.

The *jue* was always placed in the grave with the *gu [ku]* and was one of the commonest cups under the Shang. It grew rarer at the beginning of the Zhou dynasty and disappeared in about the tenth century B.C.

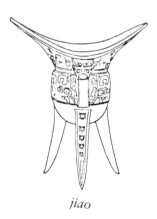

jiao

Jiao [Chiao]

A three-legged vessel similar to the *jue* with two lips exactly opposite one another and lacking the two columns. It usually had a cover.

Jiao belong exclusively to the Shang period.

jia

Jia [Chia]

This vessel is very similar to the *jue* but is larger, and its form is coarser; it has neither lip nor extension. Its legs are identical with those of the *li*.

There are square *jia* and knife-blade *jia*, but the latter shape is very rare. These vessels may be of enormous size, sometimes reaching 83 centimetres in height, and occasionally they have covers.

Their first appearance was at Zhengzhou, where they had flat bases and a body that narrowed above a shallow belly; they disappeared towards the middle of the eleventh century B.C.

he   *he*

He [Ho]

A kind of pouring pot or kettle with a cover; it had a cylindrical spout, a handle and either three or four legs.

It made its appearance towards the middle of the Shang (fourteenth century B.C.) and remained in use until the Period of the Warring States and perhaps even until the beginning of the Han dynasty.

Two quite distinct types of *he* of different dates derived from this shape. The first and earliest had a bulbous body resting on three legs identical with those of the *li*; it is known as a *lihe [li-ho]*. It had a straight spout, a semi-circular handle and a domed cover that was often attached to the body of the vessel by a little chain.

The second type of *he* made its appearance in about the fifth century B.C.; it usually had an oblong body resting on three or four short legs and might be shaped like a bird or some other animal. This variant is known as *zhui [chui]*.

Vessels for Holding and for Drinking Wine

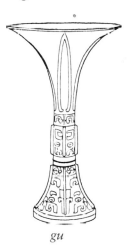

gu

Gu [Ku]

A goblet for libations of wine, it is widely flaring at the base and mouth, has a cylindrical swelling at the middle and sometimes has four flanges on the sides. Occasionally it is square and then it is known as a square *gu* or *fang gu*.

This vessel belongs exclusively to the Shang period (from the fifteenth century B.C.) and the beginning of the Zhou period. It seems to have disappeared in about the tenth century B.C.

Zun [Tsun]

This term is used for vessels of various shapes:
a) Tall vessels with a swelling in the middle and flared rims like those of the *gu* but of more massive proportions. However, there is no hard and fast dividing-line between the *gu* and the *zun*. Karlgren considers them to be two variants of the same basic form.

This variant seems to have developed in about the fourteenth century B.C.

zun   zun

the thirteenth century B.C. and was somewhat modified under the Zhou when the sides began to bulge and handles appeared on the narrow ends. This shape was abandoned at the beginning of the Period of the Spring and Autumn Annals (eighth century B.C.).

zhi

b) Large vessels, sometimes 85 centimetres in height, with a broad shoulder, a neck and a splayed foot. They are found towards the end of the Shang dynasty, in about the twelfth to eleventh century B.C.

c) This term also includes vessels in the shapes of animals (elephants, buffalos, rhinoceros, rams) or birds, named respectively *xizun [hsi-tsun]* and *niao shouzun [niao-shou-tsun]*. But in the classical texts elephant-shaped vessels are called *xiangzun [hsiang-tsun]* and those in the shape of a tiger are *huzun [hu-tsun]*. These objects were used between the eleventh century and the ninth centuries B.C.

d) Loehr places a variety of vessels dating from the Middle Shang period (fifteenth-fourteenth century B.C.) of the same type as the vessel L. 1973-71-1, displayed at the Metropolitan Museum of Art, New York, in this category. But scholars disagree over the name of the vessel, referring to it as *zun, lei* or *hu*.

Zhi [Chih] or Duan [Tuan]

A wine goblet with a flared neck, a swelling body and a hollow ring foot. Examples are found with and without cover. In some instances it resembles a sort of small *hu* without handles but with a cover. In other cases, especially when it is large, some scholars consider the *zhi* to be a variant of the *zun*.

The *zhi* made its appearance in about the middle of the Zhou dynasty (ninth-eighth century B.C.).

fang yi

you

Fang Yi [Fang-i]

A rectangular vessel with a cover, perhaps representing a primitive dwelling-place. It was used to hold food from

You [Yu]

This jar for holding and carrying wine has a large belly, which may be round or oval, a cover and a movable handle; the handle is attached at two points that are often decorated with an animal head. This vessel is easily

202

recognizable in all its variants, for it is in essence a covered jug with a movable handle.

This shape was commoner under the Shang (from the thirteenth century B.C.) than under the Zhou and disappeared in the ninth century B.C.

guang

Guang [Kuang] or Sigong [Ssu-kung]

This type of bronze vessel has no known pottery equivalent; it resembles a sauce-boat with a long cover representing the back and head of an animal (tiger, buffalo, elephant, etc.). It is a vessel for pouring and has a a semi-circular handle.

This shape is characteristic of the late Shang (end of the Anyang phase, twelfth century-eleventh century B.C.) and disappeared at the beginning of the Zhou towards the end of the eleventh century B.C.

Hu

A vase or jar, usually of considerable size, with a large body that narrows below the neck and a ring foot. The commonest shape is that of a bottle or gourd, with a low belly and an elongated neck, having rings on the sides

hu   *hu*

hu   *hu*

from which a chain may be suspended. The square form is less common and is called *fang hu*.

The *hu* made its appearance towards the end of the Shang (twelfth century B.C.) and remained in use to the end of the Zhou and under the Han dynasty. Some of the variants are:

a) The *bianhu [pien-hu]*; this is a *hu* with a body shaped like a 'flattened egg' which evolved from the Period of the Warring States onwards.

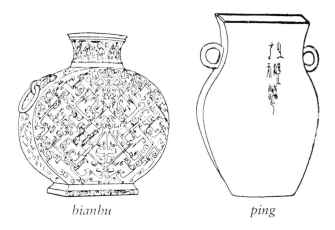

bianhu   *ping*

b) The *ping* is a *hu* with a flattened base in place of the ring foot.

Lei

A large oval-shaped vessel, with a narrow neck and sometimes a domed cover. The vessel is of the *hu* type, and it can hold a considerable quantity of liquid. The texts of the rituals inform us that the *lei* is a vessel for holding water or wine; it has high shoulders and a belly that became increasingly low. These vessels are sometimes confused with the *bu [pu]*.

203

lei

The shape made its appearance in about the thirteenth century B.C. and was abandoned in the third century B.C.

Under the Shang, the *lei* was often square; the square form is called *fang lei*.

bu

bu

Bu [Pu] or Pou [P'ou]

This is a large round jug or urn that has a comparatively narrow mouth and stands on a ring foot. Although it is usually classified as a vessel for holding wine, certain scholars believe that it was also used for keeping ground meat or offerings of grain.

This shape is often confused with the *lei*; it remained in use from the thirteenth to the tenth century B.C.

shao

Shao or Shuo

This dipper for serving liquids usually consists of a long handle with a deep bowl at the end. It is rather like a European pipe.

It was used mainly under the Shang and was radically changed under the Warring States, when a ring foot was introduced to support the bowl in which the water or wine was held.

Water Vessels

Pan [P'an]

pan

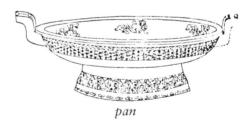

pan

According to the *Yili [I-li]* and many other classical texts, the *pan* was used for 'washing hands during ceremonies'.

This large, rather shallow, basin standing on a ring foot was used under the Shang towards the end of the fifteenth or at the beginning of the fourteenth century B.C. During the early Zhou dynasty it acquired two handles. During the ninth or eighth century B.C. the ringed foot was replaced by three feet, so that the vessel resembled a large, very flat *ding*. This shape was abandoned at the beginning of the Warring States, in about the fifth century B.C.

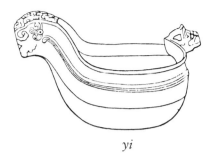

yi

Yi [I]

The variants of this ewer are extremely numerous, but in its commonest form it resembles a large bowl with a semicircular handle, a broad lip and three or four feet.

Many scholars consider that the *yi* was used to pour water into the *pan*; but the *Zuozhuan [Tso-chuan]* specifies that the *yi* was used for washing hands in certain ceremonies.

According to Mizuno this type of vessel made its appearance under the Zhou in about the eighth century B.C. and disappeared in about the fifth to fourth century B.C. But by the fifth century B.C. the feet had disappeared and were replaced by a flat base.

jian

Jian [Chien]

These huge and fairly deep basins look like a much larger version of the *pan*. They are found with or without a ring foot. These vessels are among the largest of all Chinese bronzes (an example measuring more than 1 meter in diameter is preserved in the Musée Cernuschi in Paris), and they had a short life coinciding with the Periods of the Spring and Autumn Annals and the Warring States.

Musical Instruments

Nao

A rather large bell of elliptical section with a long cylindrical handle that serves as a foot. These bells have no clapper and the sound is produced by striking them. This variety was confined to the Anyang phase of the Shang dynasty.

Zhong [Chung]

These bells of elliptical section with a circular handle were made in groups of nine to form a chime.

They made their appearance under the Zhou, and the earliest known examples would seem to be those found in 1955 in the grave of the Marquis of Cai.

Bells of the *zhong* type with a handle in the form of an openwork dragon are known as *bo [po]*.

Zheng [Cheng] and Goudiao [Kou-tiao]

These are two types of rather large bell, with a body resembling a flattened cylinder and a long handle that often terminates in a knob.

They were used principally in the Periods of the Spring and Autumn Annals and the Warring States.

Gu [Ku]

Zheng Dekun [Cheng Te-k'un] says that these drums were used from the end of the Shang dynasty onwards. At that early period the instrument took the form of a barrel standing on four short legs, but there is as yet only one documented drum of this type outside of China. This unique piece measures 79.4 centimetres in height and is preserved in the Sumitomo Collection in Kyoto. Another drum of this type, dated to the fifteen or fourteenth century B.C., was exhumed at Chongyang (near Xi'an) in 1977.

From the Zhou dynasty and until the Period of the Warring States, the shape of these drums resembled that of the rain-drums of South-east Asia.

BRONZE CASTING IN THE ANCIENT PERIOD

Method of Casting

It had been generally accepted before the 1930s that the earliest Chinese bronzes were cast by the *cire-perdue* ('lost wax' or 'wax resist') method. By this method a core of fire-proof clay of the same volume as the required vessel was encased in a model of wax. The ornament was either engraved by hand or pressed by means of a matrix on to the wax. The wax model was coated with slip containing a heat-proof substance and then with many layers of clay forming a mould or casing round the vessel. The wax melted during firing and escaped through special vents, leaving a space between the core and the casing. The liquid bronze was poured into this space. Once the bronze had cooled and the mould had been broken, the protrusions of metal were removed.

In the course of scientific excavations at Anyang and later at Zhengzhou, moulds made of grey fired clay were discovered, and it can now be stated that the Shang used the direct casting (piece-mould) method. While the moulds for weapons and agricultural implements were single or in two parts, those for ritual vessels were composite, i.e. in four or more parts. Thus, in the case of a *jue,* the body of the vessel was cast in a four-part mould locked together by tenons; other moulds were used for the legs and caps. Feet and handles were sometimes cast separately.

Preparation of the Mould

Thanks to the work of the Academia Sinica at Taipei, and more particularly of Professor Li Ji [Li Chi], it is possible to reconstruct the stages in the casting of a vessel:

1 The first and perhaps the most important stage as regards the beauty of the finished object, was to make the clay matrix.
2 The finished matrix was worked on in negative by the application of thin slabs of clay about 15 millimetres thick. After drying and firing, the negative was separated into the pieces of the mould.
3 The moulds of the outside were assembled and trued up in a frame filled with sand.
4 The inner core was adjusted so as to leave a space of between 5 and 15 millimetres between the core and the moulds.

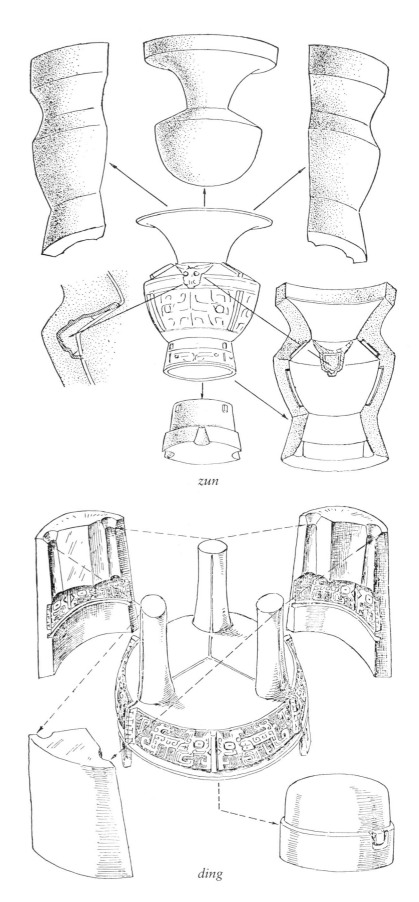

zun

ding

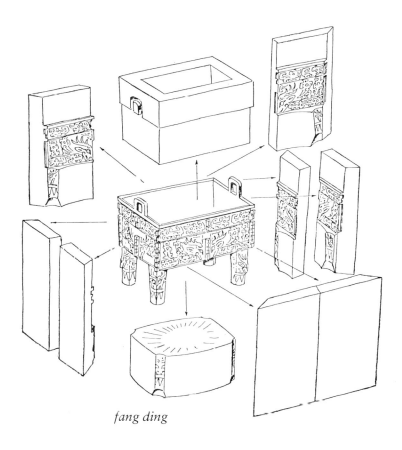

fang ding

xian

5 The liquid bronze was then poured into the space between the core and the moulds.

Traces of the joints can be seen if the bronze is examined minutely, but they are almost imperceptible. The vertical flanges or ribs which occur on many vessels follow the line of the joints.

Among the pieces recovered at Anyang were many moulds which can be identified with the vessels excavated at the same site. These different elements prove that under the Shang the great majority of bronzes were cast by the piece-mould method. It would appear that the *cire-perdue* method was not used until the middle or end of the Anyang phase.

Development of the Method of Casting

By studying the technique of casting and the different moulds, it is possible to gain some idea of the evolution of the various types of Shang bronzes. Indeed, the excavations at Anyang between 1928 and 1938 brought 176 vessels to light. These break down into forty-two *gu*, forty-two *jue*, sixteen *jia*, twenty-three *ding* and fifty-three other bronzes representing eighteen different types of vessels (*lei, pan, zun, fang yi, yu,* etc.). By studying these 176 vessels and the different moulds, Li Ji was able to conclude that the evolution of the different vessels was closely connected with the level of progress in the technology of bronze casting.

In order to determine this evolution, the 176 vessels were divided into several groups: tripods, flat-bottomed vessels, vessels with rounded bases, etc.

Development of Tripods at Anyang
(ding, jia, jue, xian, he)
a) The *ding* with Y-shaped joints was presumably the simplest type and the first to appear.
b) Circular joints came into use later. This technique is essential for tripods with 'bulbous' legs like the *xian* and the *liding*. This technique was applied later to the *ding*.
c) The *jia* inherited the method of Y-shaped joints. But its caps were developed from the technique by which the handles of the *ding* were made. Thus this was an innovation in the method of preparing the mould.
d) The *jue*, which is similar in style to the *jia*, made its appearance after the *jia*. Four moulds were used for the body and a special mould to cast the legs and the base of the vessel. (The technique by which the handles and legs were cast was more advanced.) The mould for the legs

would appear to have been influenced by the 'circular joint'.

e) The *he* was the last of this group to make its appearance.

f) The square *ding* called for an even more elaborate method.

On the basis of these six points, the classification of the Anyang tripods may be summarized as follows:

ding with Y-joints {
 liding, ding with circular joints
 jia, jue, he, fang ding
}

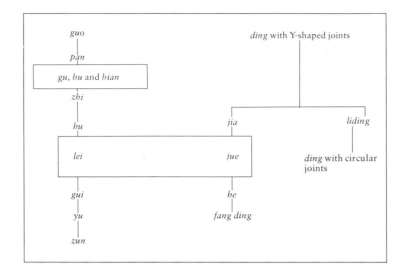

Vessels with Ring Feet and Flat Bases

There was no great difference between these two varieties of objects as regards the way in which the moulds were assembled. They were usually cast from two or three pieces divided vertically into equal parts. The evolution of these types has been reconstructed by examining accessories such as handles, relief ornament and ears.

Vessels in this category are classified as follows: *guo, pan, bu* and *bian, zhi, hu, lei, gui, yu, zu*. However, the *gu* and the *fang yi* are not included in this classification.

The *guo* is a flat-bottomed vessel and was probably cast using a single core.

Since the *pan* had an additional core for its ring foot, it has been placed after the *guo*.

Two important points emerge from this examination: sloping handles would never have been produced if two cores had not been used or if the method of composite (piece) moulds had not been known.

Chronological Order in Which Vessels Appeared

By arguing from the particular to the general, Li Ji reaches the following conclusions:

a) The *guo* was the first of the flat-bottomed vessels to appear.

b) All the vessels with round bases and a ring foot came later than the *guo*.

c) The *lei* and the *jue* appeared at the same time.

d) The *gu* precedes the *lei* and is contemporary with the *bu*.

Li Ji's important study provides an outline of the chronological order in which the shapes of ritual vessels made their appearance. It is summarized in the following table:

a
Fragment of a mould for a ritual vessel, with a *kui* dragon against a background of *leiwen*.
Grey pottery
Shang dynasty, Anyang period, thirteenth-eleventh century B.C.
Private collection

b
One half of a mould for a *zhong* bell decorated with bosses and geometric designs.
Grey pottery
Zhou dynasty, tenth-eighth century B.C.
Private collection

c
Mould for a bronze mirror bearing an inscription and an intaglio pattern of nipples.
Grey pottery
Han dynasty, second century B.C. – second century A.D.
Diameter 12 cm.
Private collection

d
Matrix for a *yue* axe with a *taotie* mask on the tang and a stylized owl on the blade.
Grey pottery
Shang dynasty, end of the Anyang period, twelfth-eleventh century B.C.
Height 23 cm.
Private collection

a

c

b

d

209

COLLECTING CHINESE BRONZES

The Western Discovery of Chinese Bronzes

In the preface to his *Bronzes archaiques chinois au Musée Cernuschi,* Vadime Elisseeff gives a brief account of the way in which Henri Cernuschi built up his collection and introduced it to the public: 'at the beginning of 1873, Cernuschi entrusted his collection to the commissioner in charge of the exhibition held at the International Congress of Orientalists at the Palais de l'Industrie;... The numerous and varied pieces excited the enthusiasm of the public at large and in particular that of the Parisian bronze-founders Barbédienne and Christofle.' The press of the day was loud in its praise of these pieces, then virtually unknown in Europe, and reflected the enthusiasm created by this important exhibition. *Le Siècle* of 7 September 1873 commented: 'Words cannot describe M. Cernuschi's collection. It contains fifteen hundred articles in bronze, more than all the museums of Europe put together possess.' The most important achievement of this first International Congress of Orientalists was to have introduced the art of the Chinese and Japanese bronze-founders to the Parisian public. Interest increased considerably with the Anyang finds, the Berlin exhibition of 1929 and the International Exhibition of Chinese Art of 1935 and 1936 in London. More recently, the exhibitions of archaeological finds of the People's Republic of China held in Paris, London, Brussels, Japan and New York gave thousands of people the opportunity to discover and appreciate this major Chinese art.

Contemporary Trends

In his preface to this book, Michel Beurdeley shows how the great collections of the first half of the twentieth century were amassed. Nowadays the rarity and high price of Chinese bronzes (especially those of the ancient period) considerably increase a collector's difficulties. There are, however, two types of collector.

Pure archaeology is the guiding principle of the first type; these collectors look for 'evidence' of a period and style. They are often content with second-rate pieces in fair condition. They hope to assemble as many pieces of different shapes and uses as possible, so as to gain an overall idea of what was made in their chosen period. Chinese bronzes of this type are comparatively easy to

find, and prices are still reasonable. But it is unfortunate that in these circumstances quality must go by the board, for an undecorated Shang or Zhou bronze, with a patina of middling quality, is usually a sorry and uninteresting object.

The second type of collector, even when of limited means, looks for quality and aesthetic beauty before all else. There are many more problems in store for him: prices will be high and objects rare, which will make his choice and final decision difficult. Collectors of this type are most particular about the quality of the casting, the forcefulness of the ornament, the perfection of the patina and the general condition of the piece. 'Purist' collectors of ancient bronzes, will often look only at vessels made under the Shang and the early Zhou dynasties. This attitude enables them to sacrifice, i.e. to re-sell, a piece that has become of secondary importance so as to buy a superior piece and thus improve their collection.

Collections of these two types are built up on entirely different premises. Both are interesting, but the second category seems to be becoming increasingly common. This opinion is echoed by dealers and auctioneers, who find that the demand for pieces of high quality is growing quite considerably, whereas prices of second-rate pieces are stationary. Artistically fine pieces are moving normally and sell comparatively easily. This change in collectors' tastes may be one of the consequences of the many exhibitions and of a better-informed public, or it may be the result of financial investment in works of art.

Where to Find Chinese Bronzes

Ancient Chinese bronzes and Buddhist gilded bronzes have become very rare in the art market since the Chinese government prohibited the export of objects over one hundred years old. The only objects now in museums and private collections are the products of private digs and pillaged tombs that left China before the years 1949 to 1950. Since the number of pieces in circulation is being steadily reduced as they are bought by or donated to museums, it is becoming increasingly difficult to find dealers or auctioneers with Chinese bronzes for sale.

Minor or second-rate pieces (undecorated vessels, vessels in poor condition, heavily restored, given a new patina and small gilded bronzes of no artistic quality) are easy to find in dealers' shops or at public auctions in England, France and America. There are very few dealers from whom one can buy major pieces of the finest quality, in perfect condition, with iconographically strong ornament. Public auctions that include such pieces are equally infrequent (on the average one or two a year in London, Paris and New York). During the 1950s it was a common occurrence to find auction sales offering over fifty ancient bronzes of very good quality. Nowadays sales including more than ten ancient vessels are comparatively rare and are an event for collectors and dealers from all over the world. Such a sale was held in London on 30 March 1978 when Sotheby Parke Bernet & Co. sold bronzes from the collection of Dr A.F. Philips; another took place in Paris on the 20 April 1980 under the direction of C. Boisgirard and A. de Heeckeren (Collection of Baron X).

Dealers are in a similar position. At one time they were able to offer a choice; there were a fair number of pieces of various types on show. The collector could compare and select the one that most appealed to him. Nowadays a dealer, whether in London, Paris or New York, is very rarely able to show a customer more than two or three pieces, and these are often of comparatively modest quality. As regards bronzes of high quality, the exhibitions held by Eskenazi Ltd. of London are the only ones at which 'purists' may admire the finest creations of Chinese bronze-founders. In addition to this illustrious dealer, who must be considered one of the greatest in the field, there is Tai of New York, who possesses a very important collection of ancient bronzes. There are others, but they operate at a less ambitious level. They include, in London, Barling Ltd., Bluett & Sons, Sydney L. Moss Ltd., John Sparks Ltd. and Spink & Son Ltd.; in Paris, M. Beurdeley, C.T. Loo and Moreau-Gobard; in New York, F. Caro and Rare Art; and G. Cröes, recently established in Brussels.

A STANDARD COLLECTION

A basic collection of Chinese bronzes need not contain more than a modest number of pieces of good quality and would include a *gu, jue, gui, ding, li, liding, zun, zhi* and *hu*, these being the commonest shapes and the most readily available on the market at relatively reasonable prices.

The collector can add to this nucleus later by acquiring the rarer shapes of vessel such as a *fang ding*, knife-blade *ding, jia, jiao, fang yi, pan, bu, you, xian* and *guang*. Since ornament varies greatly it can sometimes be interesting to acquire several pieces of similar shape, for, with one example of each type, a collector cannot expect to create a representative collection of the art of the Chinese bronze-founder.

From the point of view of quality, it is often preferable to acquire a piece of aesthetic merit, with a strong iconographic motif, a patina that pleases the eye and in perfect condition. Pieces that have been restored or have had the patina replaced should be avoided at all costs — except in the case of pieces of the utmost rarity.

NOTES FOR COLLECTORS

Several points have to be taken into account in identifying ancient Chinese bronzes. They are a) their physical nature, i.e. the nature of the metal, method of casting, corrosion and patina; b) their stylistic character, i.e. morphology and ornament in the light of the iconographic repertoires of the different periods; c) the archaeological circumstances or provenance of the object. Unfortunately this last type of information is often not forthcoming for pieces in the art trade. In such cases the pedigree of the bronze is interesting but cannot be considered a deciding factor in determining the authenticity of the piece.

Scientific Methods

Several scientific methods of dating a piece with some precision are available. The radio-carbon and thermoluminescence techniques are currently considered to be the most reliable.

Radiocarbon Dating

Based on the law of radioactive decay, this technique is the principal scientific method of dating available to the archaeologist. Since the atomic structure of isotopes is unstable, the time they take to disintegrate — their 'period of radioactivity' — varies. In the case of radiocarbon, study is based on the life of carbon-14, which is the only radioactive isotope of carbon. This method, which can be applied to all carbonaceous matter, whatever its chemical nature, i.e. to all organic matter, whether vegetable or animal, is of great value in the study of prehistory (between 30,000 and 2000 B.C.). It has no relevance to the dating of Chinese bronzes.

Thermoluminescence

The stock of energy accumulated in the minerals (quartz and feldspath) occurring in the composition of ancient ceramics may be studied by this technique. Since fired minerals and ceramics will have 'memorized' the traces of their last heating to a high temperature, thermoluminescence (i.e. the study of the emission of light produced when the temperature of a mineral that has been previously irradiated is raised) enables the technician to determine the length of time that has elapsed since that event.

This technique, perfected for dating ceramics and pottery, can be applied to the clay core of some bronzes (such as the Thai bronzes that were hollow-cast on a clay core) but not to the ancient Chinese bronzes for which direct casting in piece moulds was employed.

Although highly developed, this technique has not yet reached an acceptable level of reliability. Many uncertainties remain; these may be due to the sample having not been fired to a high enough temperature, to its having been fortuitously exposed to a later source of heat, to its having been irradiated during the taking of an X-ray photograph, to the presence of too high a proportion of calcium or feldspath. These uncertainties make it impossible to dismiss a piece as late or a fake if the thermoluminescence test is negative.

There is at the present time no scientific method of dating bronzes, except for some pieces that were hollow-cast with a clay core. Chemical analysis of the composition of the metal is the only means by which some idea of its date may be gained. Thus the presence of aluminium and zinc may sometimes be determinants in detecting a recent piece. However, the results of physical and chemical examinations are sometimes problematical and offer no possibility of precision-dating.

Fakes

Like other works of art and collectors' pieces, Chinese bronzes have been faked with varying degrees of success. From the Song period onwards, copies with anomalous ornament have been made from ancient prototypes. Fakes made deliberately to deceive collectors were produced in China during the early years of the present century, mainly during the 1930s. But few of those will stand up to close examination, and they rarely deceive a well-informed eye. Often they are aftercasts, too light in weight, and the ornament is blurred and limp. This last effect is very obvious in *leiwen* scrolls and in decorative motifs in middle relief. In other cases fakers have made bad errors of iconography; for example, a *gui* that has typical Shang ornament on the body but ears copied from a Zhou *gui* of the eighth century B.C.; and *jue* No. 11–39 in the Freer Gallery of Art, (Washington, D.C.) on which the band of ornament on the neck is in the style of the Shang bronzes of Zhengzhou, while the *taotie* masks in low relief on an undecorated ground are in the style of the Zhou vessels of the eleventh to tenth century B.C. One of the faces of the *taotie* on this *jue* is incomplete (a spur is missing), and traces of moulding are visible.

In general, the marks of a fake are inferior casting, morphological features that are similar but have minor faults and modifications (such as ring feet with sharp edges) and traces of two-piece moulding.

Patina

The patina may play a crucial part in the detection of a copy, since the patina is the result of the action on the metal of the mineral salts in the soil. Appearances and colours of patinas vary. A red patina is simply cuprite or oxidule that has taken several centuries to form. A green patina is known as malachite, which is a basic carbonate of copper. A blue patina is also a carbonate of copper and is known as azurite.

Artificial patinas are made by using acids (but the resulting patina is irregular and eats deeply into the bronze), or by applying green, red or blue pigments to the surface of the bronze or, occasionally, by painting the bronze. But simulated patina can often be detected with the aid of a magnifying glass and can be removed with alcohol or acetone.

It should, however, be noted that the presence of an artificial patina is no proof that a bronze is a fake: many bronzes have been cleaned and the patina removed.

Bronze Disease

Bronze disease is caused by chlorine that penetrates the metal. When chlorine comes into contact with copper, cuprous chloride, a chemically unstable material, is formed. When a bronze is cleaned, the cuprous chloride comes into direct contact with the humidity of the air. The humidity precipitates a chemical reaction in which the cuprous chloride is transformed into cupric chloride, which is chemically stable but powdery. This change may also come about as a result of an abrupt change of humidity, even when the bronze has not been cleaned. The disease materializes on the surface of a bronze in the form of dots of varying size, of very bright green, somewhat resembling a green powdery mould. In its active phase the disease creates pits that may in the end destroy the bronze.

There are many different forms of treatment depending on the stage to which the disease has developed. It may be treated by electrolysis, but the major disadvantage of this is that it removes the patina from the bronze, which then becomes blackish in colour. It may be treated by acids, but only if the disease is in its infancy. The affected area may be varnished or coated with paint in order to isolate the cupric chloride from the atmosphere. In this case, the disease is not treated and remains latent. The object may be placed in a vacuum in a hermetically sealed container, or the object may be preserved in a case with a nitrogenous atmosphere to preclude any chemical reaction.

Although bronze disease is common, it very rarely reaches a critical phase; however, it is advisable to keep bronzes in a dry place, to avoid sudden changes of humidity, to handle them as little as possible with bare hands and to examine them regularly for the first signs of 'disease'.

Inscriptions

Inscriptions are often a means of detecting fakes, because the fakers have either copied characters from other bronzes or invented inscriptions. Many can be recognized from stylistic or grammatical errors or from the incoherence of the text. Nevertheless, there is no need to consider every piece with a fake inscription as itself 'wrong', since characters were inscribed on many bronzes — by engraving or the use of acids — at the beginning of the present century.

Conclusion

The problem of the authenticity of an ancient Chinese bronze is more complex than would appear from these few lines. Often, however, with an experienced eye and a sound knowledge of styles, patinas and inscriptions, fakes can be detected quickly. As a last resort, chemical analysis may be a determining factor. Since fakes or late pieces are so numerous, a collector intending to acquire a bronze should first examine it closely so as to detect all its negative points before drawing his own conclusions.

1 *Liding*, the belly decorated with three *taotie* masks in high relief on a ground of *leiwen* and with a band of cicadas and whorled circles.
Shang dynasty, Anyang period, thirteenth-eleventh century B.C.
Height 20 cm.; diameter 8 cm.
Museum für Ostasiatische Kunst, Cologne

2 *Liding*, decorated in similar fashion to No. 1, with *taotie* masks and a band of cicadas.
Shang dynasty, Anyang period, thirteenth-eleventh century B.C.
Ashmolean Museum, Oxford

3 *Liding*, decorated with three *taotie* masks of bovine type and a band of snakes.
Shang dynasty, Anyang period, thirteenth-eleventh century B.C.
Museum of Far Eastern Antiquities, Stockholm

4 *Ding*, the belly decorated with *kui* dragons and triangles containing stylized cicadas; the legs decorated with geometric knife-blade motifs.
Shang dynasty, Anyang period, thirteenth-eleventh century B.C.
Height 23 cm.; diameter 15.2 cm.
Museum für Ostasiatische Kunst, Cologne

5 *Ding* with flanges; the decoration on the belly is composed of pairs of confronted *kui* dragons forming *taotie* masks in low relief against a background of *leiwen*.
Shang dynasty, Anyang period, thirteenth-eleventh century B.C.
Height 24.2 cm.
Museum für Ostasiatische Kunst, Cologne

6 *Fang ding* with flanges, each side decorated with a band of *kui* dragons and a *taotie* mask on a ground of *leiwen*.
Shang dynasty, twelfth-eleventh century B.C.
Height 22.5 cm.
Private collection

215

7   *Fang ding* with knife-blade legs; the belly decorated with pairs of confronted *kui* dragons; the legs are shaped like birds. The shape of this vessel is very rare and probably dates to the end of the Shang dynasty (eleventh century B.C.). Height 22.5 cm.
Eskenazi Ltd., London

8   *Ding* with knife-blade legs; the three legs are shaped like *kui* dragons; two *taotie* masks on the belly.
Shang dynasty, thirteenth-eleventh century B.C.
Musée Guimet, Paris

9   *Li,* decorated on the three swellings that form the body of the vessel with confronted *kui* dragons forming *taotie* masks with protruding eyes.
Shang dynasty, thirteenth-eleventh century B.C.
British Museum, London

10   *Li,* the upper part decorated with a band of eyes and spirals.
Shang dynasty, twelfth-eleventh century B.C.
Eskenazi Ltd., London

11   *Fang jia* or square *jia;* the belly is decorated in the classic manner: *taotie* masks and *kui* dragons on a ground of *leiwen;* the cover is decorated with silkworms and surmounted by a kind of horned bird in full relief.
Shang dynasty, twelfth-eleventh century B.C.
Height 22.5 cm.
British Museum, London

12   *Zhi* or *hu* with cover; scholars differ as to the designation of this vessel which is decorated with meanders and *kui* dragons.
Shang dynasty, thirteenth-eleventh century B.C.
Height 17.5 cm.
Museum für Ostasiatische Kunst, Cologne

13 *Zhi* with cover, decorated with stylized cicadas, the heads of which clearly follow the pattern of the *taotie*.
Shang dynasty, thirteenth-eleventh century B.C.
Height 19 cm.
Museum für Ostasiatische Kunst, Cologne

14 *Zhi* or *hu*, the neck decorated with a band of *kui* dragons on a ground of *leiwen*.
Shang dynasty, twelfth-eleventh century B.C.
Height 24.8 cm.
Private collection

15 *Fang yi*, the cover and each side decorated with a double-faced owl on a ground of *leiwen*.
End of the Shang dynasty, eleventh century B.C.
Height 22.5 cm.; width 16.6 cm.
Museum für Ostasiatische Kunst, Cologne

16 *Fang yi*, cover and body at the level of the flanges decorated with confronted dragons forming *taotie* masks with eyes in low relief; bands of confronted dragons on neck and foot.
Shang dynasty, Anyang period, thirteenth-eleventh century B.C.
Height 26.5 cm.; width 13.3 cm.
Museum für Ostasiatische Kunst, Cologne

17 *Gui* without handles, with an all-over design: blades or stylized cicadas, *kui* dragons, dissolved *taotie* masks and, on the foot, *kui* dragons with retroverted heads.
End of the Shang dynasty, eleventh century B.C.
Height 13.5 cm.
Museum van Aziatische Kunst, Amsterdam

18 *Gui* without handles. Together with owls and human figures, the elephants which constitute the main decorative elements are among the rarest of Shang motifs.
Shang dynasty, Anyang period, thirteenth-eleventh century B.C.
Height 13.4 cm.; diameter 21 cm.
Museum für Ostasiatische Kunst, Cologne

19 *Xian*. The upper cup is encircled by a light band of *taotie*; the three feet are decorated with *taotie* of bovine type, with eyes and muzzles in high relief.
End of the Shang dynasty or beginning of the Zhou, eleventh century B.C.
Height 37 cm.
Eskenazi Ltd., London

20 *Zun* with a band of eye motifs round the neck; the belly and foot of the vessel decorated with *taotie*.
Shang dynasty, beginning of the Anyang period, thirteenth century B.C.
Victoria and Albert Museum, London

21 *Zun* with wide shoulder; shoulder decorated with confronted *kui* reptiles and bovine masks in high relief; belly and foot decorated with incised *taotie* on a ground of *leiwen*.
Shang dynasty, thirteenth-eleventh century B.C.
Height 40.5 cm.
Museum für Ostasiatische Kunst, Cologne

22 *Zun* with a pronounced shoulder; the belly is decorated with bands of *kui* elephants, geometric motifs and *taotie* masks.
Shang dynasty, twelfth-eleventh century B.C.
Eskenazi Ltd., London

23 *Zun* adorned on the foot and the central swelling by a *taotie* mask on a ground of *leiwen*.
Shang dynasty, thirteenth-eleventh century B.C.
Height 24.3 cm.
Gisèle Cröes Collection, Brussels

24 *Zun* with knife-blade motifs on the neck; these are decorated with *taotie* masks on a ground of *leiwen*; stylized flower motifs on the central swelling and *kui* dragons on the foot.
Shang dynasty, twelfth-eleventh century B.C.
Ashmolean Museum, Oxford

25 *Hu*, the belly decorated with *taotie* masks, one above the other; they are of different sizes and the horns differ in shape.
Shang dynasty, Anyang period, thirteenth-eleventh century B.C.
Height 28 cm.
British Museum, London

26 *Hu*, decorated in five registers with motifs consisting of dissolved *taotie*, *taotie* in low relief and *kui* dragons on a ground of *leiwen*.
Shang dynasty, from Anyang, thirteenth-eleventh century B.C.
Height 40 cm.
William Rockhill Nelson Gallery-Atkins Museum of Fine Art (Nelson Fund), Kansas City

27 *Fang lei* with extremely rich decoration: birds on a ground of *leiwen* on the neck; confronted *kui* dragons, whorled circles, and triangles containing *taotie* masks on the body; and *kui* dragons with coiled tails on the foot.
Shang dynasty, thirteenth-eleventh century B.C.
Height 52.5 cm.
W. R. Nelson Gallery-Atkins Museum of Fine Art (Nelson Fund), Kansas City

28 *Yu* (?), with cover, decorated with *taotie* masks on a ground of *leiwen*.
Shang dynasty, thirteenth-eleventh century B.C.
Height 10.4 cm.
Metropolitan Museum of Art, New York

29 *Gu*, the foot and central swelling decorated with dissolved *taotie* masks; on the neck is a band of silkworms below a knife-blade motif.
Shang dynasty, Anyang period, thirteenth-eleventh century B.C.
Museum van Aziatische Kunst, Amsterdam

30 *Jiao*, the neck decorated with triangles or stylized cicadas, the body with *taotie* masks on a ground of *leiwen*.
Shang dynasty, Anyang period, thirteenth-eleventh century B.C.
Ashmolean Museum, Oxford

31 *Jiao* with a handle at the side of the body surmounted by a bovine head in low relief; the body of the vessel is decorated with *taotie* masks.
Shang dynasty, twelfth-eleventh century B.C.
Private collection

32 *Jue,* the body decorated with two *taotie* masks on a ground of *leiwen;* the neck and spout decorated with cicada-wing motifs.
Shang dynasty, thirteenth-eleventh century B.C.
Height 18.7 cm
Private collection

33 *Jue* with flanges. There are *taotie* masks on a ground of *leiwen* on the body and stylized cicadas on the neck and spouts. On the handle is the head of a bovine in relief.
Shang dynasty, thirteenth-eleventh century B.C.
Height 20 cm.
Gisèle Cröes Collection, Brussels

34 *Jue,* the body decorated with *taotie* masks on a ground of *leiwen* and triangular cicada motifs.
Shang dynasty, thirteenth-eleventh century B.C.
Private collection

35 *Jue.* On a ground of spirals, the body of the vessel and the lip are decorated with *taotie* masks, of which only the eyes are visible.
Shang dynasty, thirteenth-eleventh century B.C.
Ashmolean Museum, Oxford

36 *Nao* bell, both sides decorated with a *taotie* mask with circular, protruding eyes.
Shang dynasty, thirteenth-eleventh century B.C.
Museum of Far Eastern Antiquities, Stockholm

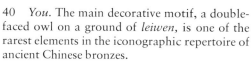

37 *Shao* or dipper for libations, two *taotie* masks adorn the bowl and the long handle is decorated with stylized cicadas on a ground of *leiwen*.
Shang dynasty, Anyang period, thirteenth-eleventh century B.C.
Length 52 cm.
A. and J. Stoclet Collection, Brussels

38 *Shao* or dipper, the handle decorated with two ovine masks in high relief and a *kui* dragon on a ground of *leiwen*.
End of the Shang or beginning of the Zhou dynasty, eleventh century B.C.
Museum van Aziatische Kunst, Amsterdam

39 *Shao* or dipper, the handle is decorated with geometric motifs of the stylized dragon type and terminates in the head of a horned monster in full relief.
Shang dynasty, Anyang period, thirteenth-eleventh century B.C.
Museum für Ostasiatische Kunst, Cologne

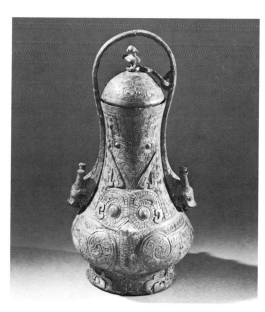

40 *You*. The main decorative motif, a double-faced owl on a ground of *leiwen*, is one of the rarest elements in the iconographic repertoire of ancient Chinese bronzes.
End of the Shang dynasty, eleventh century B.C.
Height 12.9 cm.
Museum für Ostasiatische Kunst, Cologne

41 *Guang* with cover in the shape of a horned animal. There are many iconographic elements in the very ornate decoration; they include *kui* dragons, snakes, dissolved *taotie* and elephants.
Shang dynasty, Anyang period, thirteenth-eleventh century B.C.
Asian Art Museum, San Francisco, Avery Brundage Collection

42 *Bu*, decorated in two registers: stylized dragons on a ground of spirals above, geometric motifs on the lower part of the belly.
End of the Shang dynasty, eleventh century B.C.
Height 16 cm.; diameter 16 cm.
Private collection

43 Ritual vase shaped like an elephant, decorated with *taotie* masks and *kui* dragons on a ground of *leiwen*.
Shang dynasty, thirteenth-eleventh century B.C.
Height 17.2 cm.
Freer Gallery of Art, Washington, D.C.

44 *Pan,* the ring foot is decorated with *kui* dragons of the classic type; the bowl is encircled with a band of *kui* elephants, i.e. dragons with a trunk and a coiled tail.
Shang dynasty, thirteenth-eleventh century B.C.
Private collection

45 *Fang ding* with flanges; the upper part decorated with a band of birds on a ground of *leiwen,* the lower part with studs.
Zhou dynasty, end of the eleventh century B.C.
Height 28 cm.
William Rockhill Nelson Gallery of Art—Atkins Museum of Fine Arts, Kansas City

46 *Fang ding,* decorated with studs and a band of birds on a ground of *leiwen*.
Zhou dynasty, eleventh-tenth century B.C.
Height 25.2 cm.
Museum für Ostasiatische Kunst, Cologne

47 *Ding* with flanges and a saucer-like base, decorated with a band of *taotie*.
Zhou dynasty, end of the eleventh century B.C.
Musées Royaux d'Art et d'Histoire, Brussels

48 *Gui* embellished with two dissolved *taotie* masks on a plain ground; pairs of confronted *kui* dragons adorn the ring foot.
Early Zhou dynasty, tenth century B.C.
Height 17.2 cm.; diameter 22 cm.
Gisèle Cröes Collection, Brussels

49 *Gui* without handles, or *yu,* decorated with a band of geometric motifs between borders of small circles in relief.
Zhou dynasty, eleventh-tenth century B.C.
Height 13.2 cm.; diameter 10.6 cm.
Eskenazi Ltd., London

50 *Gui* with handles, the belly decorated with *kui* dragons in the shape of elephants with coiled tails; the ring foot is decorated with a band of serpentine *kui* dragons on a ground of *leiwen.*
Zhou dynasty, eleventh-tenth century B.C.
Height 15.6 cm.
Eskenazi Ltd., London

51 *Gui* with handles and a cover, on a cubic base, the whole decorated with a wave pattern.
Zhou dynasty, ninth-eighth century B.C.
Private collection

52 *Gui,* decorated with broad horizontal grooves.
Zhou dynasty, ninth-eighth century B.C.
Height 14.5 cm.
Ashmolean Museum, Oxford

53 *Yu.* Interlaced dragons ornament the body, while the neck and foot are decorated with scale-like motifs called *linwen* and a band of stylized *kui* dragons.
Mid Zhou dynasty, eighth century B.C.
Height 15 cm.; diameter 24 cm.
Simone and Alan Hartman Collection, New York

54 *Zun* with flanges, decorated with dissolved *taotie* masks on plain grounds in two registers.
Zhou dynasty, eleventh century B.C.
Victoria and Albert Museum, London

55 *Guang* with a cover in the shape of an animal head; the body of the vessel and the cover are decorated with crested birds on grounds of *leiwen* in several registers.
Zhou dynasty, eleventh-tenth century B.C.
Height 31.5 cm.; width 36 cm.
Art Museum, Princeton University, Princeton

56 *Guang*, with a cover representing the back and head of an animal; the flanged body of the vessel is decorated with *taotie* masks and *kui* dragons on a ground of *leiwen*.
Zhou dynasty, eleventh-tenth century B.C.
Metropolitan Museum of Art (Rogers Fund 1943), New York

57 *You*, the neck and foot decorated with a band of *kui* dragons in low relief; an impressive horned *taotie* mask on the belly.
Zhou dynasty, eleventh century B.C.
Height 23 cm.
Victoria and Albert Museum, London

58 *You*, the neck and cover decorated with bands of winged *kui* dragons on a ground of *leiwen*.
Zhou dynasty, eleventh-tenth century B.C.
Height 32.5 cm.; width 20.5 cm.
Art Museum, Princeton University, Princeton

59 *You* with handles fixed to the body by animal heads in middle relief; the body and cover are decorated with a band of *kui* dragons.
Zhou dynasty, eleventh-tenth century B.C.
Height 20 cm.; width 26 cm.
Maurice Bérard Collection, Paris

60 *You*, cover and body of the vessel decorated with a band of crested birds on a ground of *leiwen*.
Zhou dynasty, eleventh-tenth century B.C.
Private collection

61 *You* with flanges. The body is adorned with a large dissolved *taotie* mask and a band of *kui* dragons; on the ring foot are pairs of confronted *kui* dragons.
Zhou dynasty, ninth–eighth century B.C.
Private collection

62 Vessel in the shape of a buffalo, with all four feet on the ground and the head turned slightly to the right.
Zhou dynasty, ninth–seventh century B.C.
Length 26 cm.
Eskenazi Ltd., London

63 An imposing *hu* with cover and two handles at the sides; the body is decorated in an all-over pattern of interlacing dragons.
Zhou dynasty, ninth–eighth century B.C.
Musée Guimet, Paris

64 *Hu* with two handles at the side; the body is decorated with spiral interlace in low relief and stylized *taotie* masks in middle relief.
Zhou dynasty, ninth–eighth century B.C.
Private collection

65 Recumbent buffalos decorated with flat curving bands and hooks thrown into relief by the incised spirals and lines that run parallel with them.
Zhou dynasty, eighth–seventh century B.C.
Height 8 cm.; width 16 cm.
Private Collection, Paris

66 Oval *dui* on three legs; the cover and base of the vessel are decorated with hunting scenes.
End of the Zhou dynasty—Period of the Warring States, fifth–third century B.C.
Height 16 cm.
Private collection

67   *Yi*, a ewer on four legs; the semi-circular handle terminates in an animal head.
Zhou dynasty, ninth-eight century B.C.
Length 36 cm.
Musée Guimet, Paris

68   Ritual vessel in the shape of an owl with movable cover.
Zhou dynasty, eleventh-tenth century B.C.
Height 22.5 cm.
Yale University Art Gallery (gift of Mrs William H. Moore for Hobart Moore and Edward Small Moore Memorial Colection), New Haven, Conn.

69   Vessel with handle in the shape of an owl.
Zhou dynasty, eleventh-tenth century B.C.
Height 23.5 cm.
Hakutsuru Art Museum, Kobe

70   *Lei* jug adorned with bands of interlaced dragons and *taotie* masks in low relief to which rings are attached.
End of the period of the Spring and Autumn Annals, sixth-fifth century B.C.
Height 29.5 cm.
Asian Art Museum, San Francisco, Avery Brundage Collection

71   *Dui* with traces of inlay.
Period of the Spring and Autumn Annals, fifth-fourth century B.C.
Fogg Art Museum, Cambridge, Mass.

72   *Hu* decorated with five bands of interlacing dragons and two *taotie* masks in low relief forming points of attachment for rings.
End of the Period of the Spring and Autumn Annals—beginning of the Period of the Warring States, sixth-fourth century B.C.
Private collection (sold C. Boisgirard and A. de Heeckeren, Paris, 1980)

73 *Hu* decorated on either side with *taotie* masks in low relief; the belly is decorated in low relief with an interlace of stylized *kui* dragons. End of the Period of the Spring and Autumn Annals—beginning of the Period of the Warring States, sixth-fifth century B.C.
Height 32 cm.
Private collection

74 *Bianhu* with cartouches of stylized interlaced dragons or spirals termed *yuzhuangwen*.
Period of the Warring States, fifth-third century B.C.
Los Angeles County Museum of Art, Los Angeles

75 *Ding* with cover, decorated with an interlace of stylized dragons; on the cover are three recumbent bovines in high relief forming points of attachment for rings.
Period of the Spring and Autumn Annals, seventh-fifth century B.C.
Private collection

76 Ritual vessel shaped like a tapir.
Period of the Spring and Autumn Annals, seventh-fifth century B.C.
British Museum, London

77 *Dou*, the body and cover are decorated with geometric T motifs.
End of the Period of the Spring and Autumn Annals—beginning of the Period of the Warring States, sixth-fifth century B.C.
Frank Caro Collection, New York.

78 *Fu* with cover, decorated with serpent interlace.
Zhou dynasty-Period of the Spring and Autumn Annals, seventh-fifth century B.C.
Courtesy of the Art Institute of Chicago

79   Kneeling figure.
Bronze
End of the Zhou dynasty-Period of the Warring States, fifth-third century B.C.
Private collection

80   *He*, kettle with a spout in the shape of a bird's head.
Period of the Warring States, fifth-third century B.C.
Musée Cernuschi, Paris

81   *Xian* decorated with *taotie* masks in low relief.
Period of the Warring States, fifth-third century B.C.
Height 40.5 cm.
Art Museum, Princeton University, Princeton, N. J.

82   Cow or other bovine animal.
End of the Period of the Warring States, fourth-third century B.C.
Height 11 cm.; width 13.5 cm.
Asian Art Museum, San Francisco, Avery Brundage Collection

83   *Dui,* ritual vessel decorated with bands of T-shaped spirals.
Period of the Warring States, fifth-fourth century B.C.
Ashmolean Museum, Oxford

84   Pear-shaped vessel on a pedestal, decorated with several registers of geometric spirals.
Period of the Warring States, fifth-third century B.C.
British Museum, London

85 *Ding* with cover, decorated with bands of stylized interlaced dragons; three rebumbent bovines in middle relief adorn the cover.
Period of the Warring States, fifth-third century B.C.
Musée Cernuschi, Paris

86 Tripod vessel shaped like a censer and ornamented with *taotie* masks in middle relief; the cover is decorated with birds in full relief.
Period of the Warring States, fifth-third century B.C.
Height 21 cm.
Metropolitan Museum of Art (Rogers Fund 1947), New York

87 *Hu* adorned with two *taotie* masks in low relief to which ring handles are attached.
Han dynasty, second century B.C. — second century A.D.
Height 43 cm.
Spaegnaers Collection, Anvers

88 *Hu* decorated with bands of geometric ornament and two *taotie* masks in low relief.
From Southern China.
Han dynasty, second century B.C. — second century A.D.
Musée Cernuschi, Paris

89 Bottle-shaped *hu* with a cover and chain handle for suspending it; decorated with two *taotie* masks in low relief.
Han dynasty, second century B.C. — second century A.D.
Height 42.5 cm.
Gisèle Cröes Collection, Brussels

90 Drum-shaped container for cowries. The cover is decorated with a scene of a sacrifice, figures of men and animals in full relief.
Dynasty of the Former Han, first century B.C.
From Tomb No. 13 at Shi Zhai shan [Shih-chai-shan], Yunnan province
Height 39.5 cm.; diameter 40 cm.

91   Vessel in the shape of a duck(?)
Bronze with a dark patina
Han dynasty, second century B.C. — second century A.D.
British Museum, London

92   Small free-standing figure of a horse.
Han dynasty, second century B.C. — second century A.D.
Height 8 cm.
Musées Royaux d'Art et d'Histoire, Brussels

93   Bronze cooking stove.
Han dynasty, second century B.C. — second century A.D.
Musées Royaux d'Art et d'Histoire, Brussels

94   *Hu* in the shape of a gourd, with a cover formed by the figure of a bird.
Han dynasty, second century B.C.—second century A.D.
Musée Cernuschi, Paris

95   *Yue* axe, with a *taotie* mask on the tang and *kui* dragons on the upper part.
Shang dynasty, thirteenth-eleventh century B.C.
Museum of Far Eastern Antiquities, Stockholm

96   *Yue* axe, with a *taotie* mask on the tang and *kui* dragons and cicadas on the upper part of the blade.
Shang dynasty, thirteenth-eleventh century B.C.
Museum of Far Eastern Antiquities, Stockholm

97 *Chu* axe; the tang decorated with a motif of interlacing serpents inlaid with turquoises.
Shang dynasty, thirteenth-eleventh century B.C.
Musée d'Etat de Mariemont, Mariemont, Belgium

98 *Qi* axe, decorated with an imposing open-work *taotie* mask and spirals.
Shang dynasty, thirteenth-eleventh B.C.
From Anyang
Height 35 cm.; width 37.8 cm.
Museum fur Östasiatische Kunst, Cologne

99 *Nao* spearhead; the lower part decorated with a *taotie* mask in low relief on a ground of spirals.
Shang dynasty, thirteenth-eleventh century B.C.
From Royal Tomb HPKM 1001, Anyang
Length 35.5 cm.
Institute of History and Philology, Academia Sinica, Taipei

100 Warrior's helmet decorated with an imposing *taotie* mask on the front.
Shang dynasty, thirteenth-eleventh century B.C.
From Royal Tomb HPKM 1004, Anyang
Institute of History and Philology, Academia Sinica, Taipei

101 Side view of the warrior's helmet in No. 100
Shang dynasty, thirteenth-eleventh century B.C.
From Royal Tomb HPKM 1004, Anyang
Institute of History and Philology, Academia Sinica, Taipei

102 *Ge* halberd; the tang decorated with a *kui* dragon.
Shang dynasty, thirteenth-eleventh century B.C.
Museum van Aziatische Kunst, Amsterdam

Glossary

abhaya mudra
Gesture of the absence of fear; right hand raised, fingers extended, palm forwards.

Akasagarbha
Bodhisattva of the Essence of the Void, the most esoteric being of Tang Buddhism. His attributes are a pearl (or jewel), a sword or the three-pointed *vajra*.

Amitabha or 'Infinite Brilliancy'
Also called Amitayus, meaning infinite longevity. This Buddha of the Western Paradise is a transcendant Buddha of the Mahayana school. Amitabha is one of the most popular jinas.

amrita
A ritual vessel containing the Elixir of Life; one of the attributes of Avalokitesvara Padmapani.

Ananda
One of the Sakyamuni Buddha's two favourite disciples.

anjali mudra
Gesture of adoration, palms of the hands pressed together at the level of the breast.

apsaras
Minor Buddhist divinities whose role somewhat resembles that of angels in Christianity. They often have feminine features, and they hold a musical instrument.

arhats
monks, disciples of the Master. They are the apostles of Buddhism. They are depicted with shaven heads, wearing the monastic habit, hands very often performing the *anjali mudra* or carrying a sacred vessel. Known as *luohans* in China.

Avalokitesvara or Lokesvara
A bodhisattva of mercy and compassion who delayed his entry into Nirvana in order to save souls. He is the greatest and most popular bodhisattva of Mahayana Buddhism. In his head-dress he wears a small image of the Amitabha Buddha, of whom he is the emanation. He is known in China by the name of Guanyin.

Avalokitesvara Cintamanicakra
Figure of Avalokitesvara holding the *cintamani* or magic jewel.

Bafang [Pa-fang]
Barbarian tribes who opposed the Shang armies, mainly during the reign of King Wu Ding.

bai [pai]
A fixed or movable grid placed inside a *xian*.

bairu [pai-ju]
Type of bronze mirror decorated with nipples or bosses joined together by curving lines or clouds.

bang [pang]
A bronze fitting used under the Shang and Zhou to reinforce longbows.

bhadrasana
Seated position in the European style: the figure sits on a throne with both legs pendent.

Bhaishajyaguru
Buddha of Medicine or 'Master of Remedies', known as Yaoshi in China. May be identified by the golden fruit *haritaki* or a medicinal jar that he holds in his hands. He is the Buddha of the East. He vowed to become as limpid as beryl and to cure all ills, diseases and distress. Authorship of a treatise on medicine is attributed to him.

bhumisparsa mudra
Gesture of taking the Earth to witness: right hand resting on the leg, palm inwards, fingers pointing to the ground. Represents the Buddha's victory over Mara, the god of the realm of desires.

bianhu [pien-hu]
A *hu* in the shape of a flattened egg which evolved in the Period of the Warring States (453–221 B.C.).

bo [po]
Counts, high officials of the kingdom; under the Shang

those who had distinguished themselves for their qualities or military prowess.

bodhisattva
A being on the way to Enlightenment, destined to attain to Buddhahood in the near future.

bu [pu]
Bronze ritual vessel resembling a spherical urn with a relatively narrow opening and a ring foot. This form evolved in the thirteenth century B.C. and disappeared in about the tenth century B.C.

Budai Heshang [Pu-t'ai Ho-shang]
'Laughing Buddha', image of prosperity and contentment; depicted as a happy, corpulent figure.

Buddha
The 'Enlightened One'. Buddhism distinguishes between Prateyka Buddhas, the Samyaksam Buddhas and the Buddhas of the Past. Transcendent Buddhas are peculiar to Mahayana Buddhism.

Buddhas of the Past
They preceded Sakyamuni, the historical Buddha. In Mahayana Buddhism they are usually seven in number, corresponding to the sun, the moon and the five planets. They are depicted in groups of three, five or seven, for other schools believe that Sakyamuni was preceded by five Buddhas associated with the five directions (the four cardinal points and the centre) and three other Buddhas.

Candraprabha
Bodhisattva of the moon. He holds a crescent moon in his hand.

caoye [ts'ao-yeh]
Type of bronze mirror decorated with a central square containing an inscription and with stylized flowers in each corner.

chen [ch'en]
Title named in the oracular inscriptions meaning feudal lord. Under the Shang the *chen* were the rank below that of prince.

Chifang [Ch'ih-fang]
Barbarian tribes mentioned in the *jiaguwen*. The Shang armies of King Wu Ding fought against them.

chu [ch'u]
Term for two types of weapon derived from the axe or halberd. The first has a V-shaped blade, the second has a haft that is rounded to the shape of the wooden socket into which it fits.

Chunqiu [Ch'un-ch'iu]
Period of the Spring and Autumn Annals.

cintamani
A magic jewel, one of the attributes of the bodhisattva Kshitigarbha.

Cintamanicakra
Depiction of the bodhisattva Avalokitesvara holding the *cintamani* in his hand.

dao [tao]
A knife; the weapon used by Shang and Zhou charioteers.

dazhuan [ta-chuan]
Type of script known as 'great seal'.

Dharmacakra
The Wheel of the Law. The Wheel or *cakra* in Buddhism is a symbol of the Buddha's teaching and of his First Sermon.

dharmacakra mudra
Gesture of setting the Wheel of the Law in motion, symbolizing the Buddha's teaching. In Mahayana Buddhism, the gesture often characterizes Vairocana, the transcendent Buddha.

dhyana mudra
Gesture of meditation, arms bent, hands resting in the lap, either one above the other or with interlocking fingers.

ding [ting]
Name of a tripod in which food was cooked. Probably the earliest shape to have been cast in bronze.

Dizang [Ti-tsang]
Chinese name of the bodhisattva Kshitigarbha.

dou [tou]
A hemispherical cup on a tall flaring foot. The cover looks like a separate cup.

duan [tuan]
Type of bronze goblet similar to the *zhi*.

duan [tuan]
Name of *gui* in the *jinwen* inscriptions.

dui [tui]
Term used from the Song onwards to denote a spherical vessel, a type of round *gui* with a cover.

duozizu [to-tzu-tsu]
Under the Shang, the *duozizu* were princes who lived in the capital and whose role was to participate in military operations or to re-establish internal order.

dvarapalas
These are the guardians or defenders of Buddhism; their emblem is the *vajra*.

Emituo Fo [E-mi-t'o Fo]
Chinese name of the Amitabha Buddha.

Erbo [Erh-po]
Barbarian tribes surrounding the Shang kingdom, against whom the armies of King Wu Ding fought.

Erya [Erh-ya]
Dictionary of the Chinese language in the form of an encyclopaedia and a series of glosses. Although its date is unknown, it was prior to the Han Empire.

erzi [erh-tzu]
Type of mirror, the sole decoration of which is an inscription, each alternate character being *er*.

fangbo [fang-po]
Title given to foreign chieftains who were conquered and then integrated into the Shang state.

fang ding [fang-ting]
Rectangular ritual vessel with four legs. It made its appearance in the fifteenth century B.C. and disappeared in about the tenth century B.C.

fang gu [fang-ku]
Square form of the vessel *gu*.

fang hu [fang-hu]
Rectangular form of the vessel *hu*.

fang yi [fang-i]
A ritual vessel of rectangular section with a cover. Made its appearance in about the thirteenth century B.C. and was abandoned in about the eighth century B.C.

Fo
Chinese term for Buddha

fu
Rectangular ritual vessel on a ring foot, with a cover of the same shape and size as the vessel itself. Typical Zhou vessel.

Fu-X
Ladies; name given either to the wives of Shang princes or, more probably, a feudal title given under the Shang to women of very high rank = Lady X.

Ganesha or Ganesa
'Lord of the *ganas*' (minor divinities), god of the knowledge that destroys obstacles. He is depicted as a fat man with the head of an elephant.

gao [kao]
Name of a Shang ceremony in honour of ancestors.

ge [ko]
A halberd, one of the two weapons of war under the Shang.

ghanta
A little bell, one of the attributes of the bodhisattva Vajrasattva.

goudiao [kou-tiao]
Type of bell used during the Periods of the Spring and Autumn Annals and the Warring States.

gu [ku]
A cup with wide, spreading base and mouth, used for wine libations. This vessel made its appearance in about the fifteenth century B.C. and disappeared in about the tenth century B.C.

gu [ku]
Bronze barrel-shaped drum.

guang [kuang]
A jug-like vessel; cover in the shape of an animal's back and head: late Shang, early Zhou shape.

guangding [kuang-ting]
Vessel of hybrid shape, having the feet of a *ding* and the contours of a *guang*.

Guanyin [Kuan-yin]
Chinese name of the bodhisattva Avalokitesvara who evolved into a feminine divinity in China in the tenth century.

gui [kuei]
Ritual vessel consisting of a circular cup on a ring foot. Exists with and without ears.

guo [kuo]
A flat-bottomed rectangular vessel. An example was unearthed at Anyang during the excavations of 1928-38.

haritaki
A golden fruit, one of the attributes of the Buddha Bhaishajyaguru, Buddha of Medicine.

he [ho]
A kettle or jug with a cover, having a cylindrical spout, a handle and three or four legs.

Hinayana
The Small Vehicle. The name of the Buddhist sects that are closest to the original tradition, especially Theravada Buddhism.

hou
A feudal title corresponding to marquis.

hu
A vessel with a wide body narrowing towards the neck and a ring foot. The commonest form under the Zhou and the Han dynasties.

hui
A ritual vessel of the same family as the *gui*, but with a cover.

huzun [hu-tsun]
The *zun* in the shape of a tiger.

jia [chia]
A tripod resembling the *jue*, but larger and clumsier in shape. A *jia* has neither spout nor extension. There are square *jia*, but they are very rare.

jiaguwen [chia-ku-wen]
The word for Shang-period inscriptions on bones and tortoise-shells; they are considered the oldest examples of written texts in China .

jian [chien]
A very large and moderately deep vessel, with or without ring foot. *Jian* are the largest bronze vessels of the Periods of the Spring and Autumn Annals and the Warring States.

jiao [chiao]
A tripod vessel with two lips for pouring positioned exactly opposite one another. This archaic shape was cast uniquely under the Shang.

jinwen [chin-wen]
The word for the writing on bronze of the Shang and Zhou periods.

jue [chüeh]
Tripod vessel with one handle at the side, a broad lip for pouring, a pointed extension and two vertical columns surmounted by a kind of cap.

Kasyapa
One of the two favourite disciples of the Sakyamuni Buddha.

khakkhara
One of the attributes of the bodhisattva Kshitigarbha, representing a crescent-shaped dagger surmounted by a crossed *vajra*; called *xiyhang [hsi-chang]* in Chinese.

Kshitigarbha
A bodhisattva known in China by the name of Dizang [Ti-tsang]. He can be identified by his shaven head, monastic habit and his two attributes: the *khakkhara* and the *cintamani*.

kui [k'uei]
A type of dragon; a very common motif in the decorative iconography of ancient Chinese bronzes. The creatures occur in a great variety of forms inspired by representations of living animals, such as tigers, elephants, rams, birds, etc.

lalitasana
Seated position of relaxation with the left leg bent and the right leg pendent.

Laozi [Lao-tzu]
Sage and thinker, father of Taoism, putative author of the *Daodejing [Tao-te-ching]*.

li
A three-legged — very rarely four-legged — cauldron consisting of hollow legs that join to form the body of the vessel.

lianzhou [lien-chou]
'Threaded beads', another name for *bairu* mirrors.

liding [li-ting]
Ritual vessel of hybrid shape with three shallow but distinct depressions above the legs.

lihe [li-ho]
A variety of the *he* with a bulbous body on three feet (identical with those of the *li*), a cylindrical spout, a semi-circular handle and a domed cover.

lei
A large ovoid vessel with a high shoulder and a belly that tapers towards the base. Under the Shang a *lei* was often square; this type is known as *fang lei*.

leiwen
Geometric motif often used as background on ancient Shang bronzes. The *leiwen* is a combination of spirals and meanders.

linwen
A form of coil that occurs in the decoration of bronze ritual vessels of the Periods of the Spring and Autumn Annals and the Warring States.

lokapalas
Guardian-protectors of the four corners of Buddhist shrines.

Lotus Sutra or *Lotus of Right Law [Saddharmapundarika Sutra]*
The most celebrated of the Mahayana sutras in China and in the Buddhist world, doubtless because of the light it throws on the Mahayana conception of the Buddhas and bodhisattvas.

luohan
Chinese word for an *arhat*. A monk following in the footsteps of the Master.

Mahasthamaprapta
Bodhisattva attendant on the Sakyamuni Buddha.

Mahayana
The 'Great Vehicle' or 'Great Means of Progression'. Buddhist doctrine in which speculation on the nature of the Buddha and the roles of the bodhisattvas are of essential importance. Mahayana Buddhism is a development of Hinayana; it is easier than the latter because it is less demanding and promises greater rewards.

Maitreya (bodhisattva)
One of the great bodhisattvas in Mahayana Buddhism. He is depicted with a stupa in his head-dress.

Maitreya (Buddha)
The next Buddha or Buddha of the future. Sakyamuni attained to Enlightenment before him.

Manjusri
Bodhisattva of wisdom, known as Wenshu in China. After Avalokitesvara, he is the most important character in the *Lotus Sutra*. He is depicted in Chinese art riding a lion or holding a book or sword in his hand.

mao
A spearhead.

Mengfang
Barbarian tribes opposed by the forces of the Shang king Di Xin [Ti Hsin].

mingguang [ming-kuang]
Name given to mirrors on which the characters *ming* and *guang* appear.

mudra
Gesture of the hands and fingers to which a mystical and magic significance is attributed.

nai
According to the *Erya*, a *nai* is a large *ding*.

nao
A bell of elliptical section with a long handle that serves as a foot. This type of bell appears only to have been made under the Shang.

niaoshouzun [niao-shou-tsun]
A bird-shaped variant of the *zun*.

Padmapani
See Avalokitesvara.

pan [p'an]
A ritual vessel in the form of a shallow basin on a ring foot. The *pan* acquired ears at the beginning of the Zhou dynasty. From the ninth or eighth century B.C., the ring foot was replaced by three feet.

panlong [p'an-lung]
Zoomorphic ornament in the form of stylized dragons.

Pilushena [P'i-lu-she-na]
Chinese name of the Vairocana Buddha.

ping [p'ing]
Ritual vessel of the *hu* type but with a flattened base in place of the ring foot.

Prabhutaratna
Buddha of the Past, much venerated under the Wei and the Sui dynasties; during these periods he was depicted seated beside Sakyamuni.

Pratyeka Buddhas
Buddhas who achieved Enlightenment for themselves alone.

Pusa [P'u-sa]
Chinese term for a bodhisattva.

Puxian [P'u-hsien]
Chinese name of the bodhisattva Samantabhadra. He is usually shown riding an elephant and symbolizes the union of the highest intelligence and good actions.

qi [ch'i]
Large axe (measuring 30−40 cm in width) with convex outline.

qingbai [ch'ing-pai]
Name given to mirrors on which the characters *qing* and *bai* appear.

Renfang [Jen-fang]
Barbarian tribes mentioned in the *jiaguwen* of the reign of the Shang king Di Xin.

Sakyamuni
Or the 'Sage of Sakya', name of the historical Buddha.

samadhi
Attitude of meditation. The figure is seated in the Indian position with his hands resting on each other.

Samantabhadra
See Puxian.

Samyaksambuddha or Sammasambuddha
Those who are 'perfectly enlightened'; they alone can instruct people. The most celebrated of them is Sakyamuni or the historical Buddha.

shao
Bronze dipper for serving liquids.

Shiji [Shih-chi]
Or *Records of the Historian Ssu-ma Ch'ien* (135?−93? B.C.), a history of China from the beginnings to the Han period by Sima Qian.

Shijing [Shih-ching]
The Book of Odes, one of the classical Chinese texts.

shiru-sihui [shih-ju ssu-hui]
Bronze mirrors decorated with four nipples and four dragons, each measuring about one quarter of the diameter of the piece.

shouzhou [shou-chou]
Very thin bronze mirrors, decorated with interlacing dragons, coils and triangles on a ground of *leiwen*.

shuo
A form of bronze dipper (*see shao*).

Shuowen [Shuo-wen]
The earliest dictionary of the Chinese language, compiled under the Han dynasty.

sigong [ssu-kung]
Another name for *guang* vessels.

Suryaprabhasa
Bodhisattva of the sun, shown holding the solar disc in his left hand.

sutra
A rule expressed in aphorisms. Sutras are treatises containing the rules of Buddhism.

TLV mirrors
Mirrors with decorative motifs resembling the letters T, L, and V.

taotie [t'ao-t'ieh]
Iconographic motif on ancient Chinese bronzes. Its significance has been the subject of many hypotheses. Its most highly evolved form consists of two *kui* dragons (or gluttons) in profile and confronted.

Theravada
Doctrine of the Elders. Represents the heritage of the Pali tradition of early Buddhism. Its basic text is the Pali Canon, considered to be the authentic teaching.

tribhanga
A 'dancing' movement or triple flexion. A very pronounced S-curve or 'swaying' posture, characteristic of Indian iconography.

ushnisha
A protruberance on the head, one of the marks of the Buddha.

Vairocana
Cosmic Buddha. He carries a solar disc on his shoulders; on it are a bird and a crescent moon, symbols of the sun and moon respectively.

vajra
Symbol of knowledge and the Buddhist force that destroys evil passions and is indestructible itself; shaped like a thunderbolt.

Vajrapani
Divine guardian, holder of the thunderbolt. Minor divinity of the Buddhist pantheon, charged with guarding the approach to temples.

vajrasana, vajraparyanka or *padmasana*
Seated posture, legs tightly crossed, soles of the two feet visible.

Vajrasattva
Bodhisattva regarded as the Essence of the Thunderbolt. His attributes are the *vajra* and the *ghanta*.

vara mudra or *varada mudra*
Gesture of giving and charity: arm extended, right hand open and out-stretched.

Vinaya
Discipline. One of the 'three baskets' *(Tripitaka)* of the Buddhist Canon; contains the body of disciplinary rules with an account of the reasons that have motivated them.

virasana
Seated position with the right leg placed on top of the left leg.

vitarka mudra
Gesture of discussion: hand half-open, thumb and index-finger touching. The gesture may be made with one or both hands.

Wenshu
Chinese name for the bodhisattva Manjusri.

xian [hsien]
Tripod vessel for steaming food. The lower part is similar to the *li;* the upper part is cup-like with a grid as base.

xiangzun [hsiang-tsun]
Zun in the shape of an elephant.

xiaozhuan [hsiao-chuan]
Script known as 'small seal', which may have been invented by the minister of the mythical First Emperor; standard script.

xingyun [hsing-yün]
'Stars and clouds'; another name for *bairu* mirrors.

xizun [hsi-tsun]
Bird-shaped vase of the *zun* type.

xu [hsü]
Vessel resembling a *fu,* i.e. rectangular in form but with rounded corners. The cover is very much smaller than the body of the vase.

yan [yen]
See xian.

Yaoshi [Yo-shi]
Chinese name for the Bhaishajyaguru Buddha.

yaxing [ya-hsing]
Clan marks in inscriptions on bronze.

yi [i]
According to the *Erya*, a *yi* is a *ding* with ears on the outside of the body.

yi [i]
A ewer resembling a large bowl with a semi-circular handle, a large spout and three or four feet.

yi ding [i-ting]
A very rare vessel of hybrid form combining the feet of a *ding* and the body of a *guang* or a *yi*.

Yili [I-li]
Book of formal behaviour; early records of ancient rituals (about the fourth-third century B.C.).

you [yu]
Jug with cover and movable handle.

yu [yü]
Ritual vessel of the *gui* type with L-shaped ears below the mouth of the vessel

yue [yüeh]
An axe with a narrow blade and concave edges, ending in a small handle.

zai [tsai]
According to the *Erya*, a *zai* is a *ding* with a narrow opening.

zeng [tseng]
Name of the upper part of a *xian*.

Zhanguo [Chan-kuo]
Period of the Warring States (453 – 221 B.C.).

zhi [chih]
Goblet with a flared neck, bulbous body and ring foot.

zhong [chung]
A bell, elliptical in section, with a circular handle.

zhui [chui]
Ritual vessel of the *he* type, with an oblong body on three or four short legs that may be in the shape of a bird or other creature.

zun [tsun]
A vessel that bulges in the middle and has flared edges resembling a *gu*.

Zuozhuan [Tso-chuan]
Traditions of Zuo. A chronicle that appeared in the form of a commentary on the *Annals of Lu*; throws light on Zhou society.

Chronological Table

Shang dynasty		(?) 17th–11th century B.C.
Zhou dynasty		11th century–256 B.C.
	Western Zhou	11th century–770 B.C.
	Eastern Zhou	770–256 B.C.
	Warring States	453–221 B.C.
Qin dynasty		221–06 B.C.
Han dynasty		206 B.C.–A.D. 220
	Western Han	206 B.C.–A.D. 8
	Eastern Han	A.D. 25–220
Three Kingdoms		220–65
Jin dynasty		265–420
Period of the Six Dynasties		420–589

North	Northern Wei		386–535
	Eastern Wei		534–50
	Western Wei and Northern Zhou		535–81
South	Song		420–79
	Southern Qi		479–502
	Liang		502–57
	Chen [Ch'en]		557–89

Sui dynasty		581–617
Tang dynasty		618–906
Five Dynasties		907–60
Song dynasty		960–1279
	Northern Song	960–1127
	Southern Song	1128–1279
Yuan dynasty (Mongolian)		1260–1368
Ming dynasty		1368–1644
Qing dynasty		1644–1912

Map of China

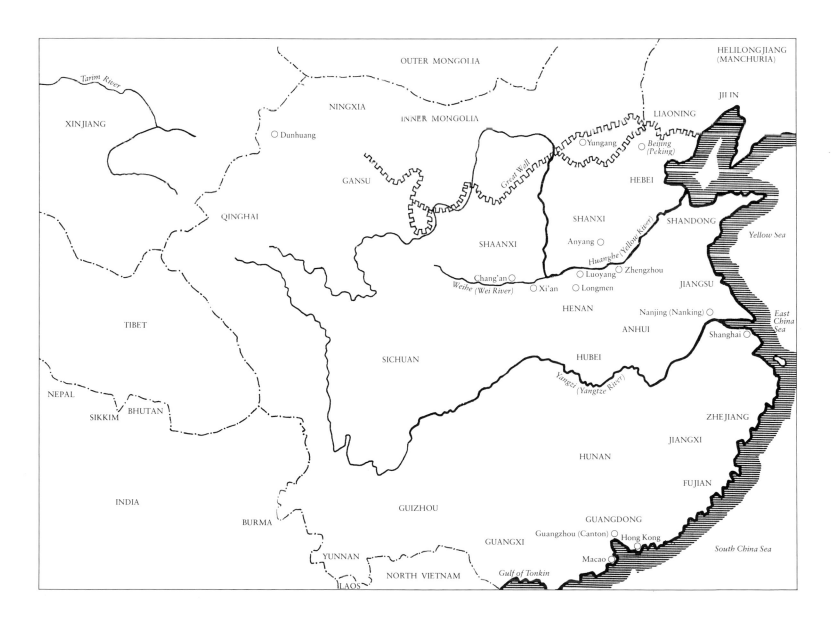

List of Museums

Australia

Australian National Gallery, Canberra

Austria

Museum für Völkerkunde, Vienna

Belgium

Musée d'Etat de Mariemont, Mariemont
Musées Royaux d'Art et d'Histoire, Brussels

Canada

Montreal Museum of Fine Arts
Royal Ontario Museum, Toronto

China

National Palace Museum, Peking

France

Musée Cernuschi, Paris
Musée Guimet, Paris

Germany

Museum für Kunst und Gewerbe (Ostasiatische Kunst), Hamburg
Museum für Ostasiatische Kunst, Berlin
Museum für Ostasiatische Kunst, Cologne

Great Britain

Ashmolean Museum of Art and Archaeology, Oxford
British Museum, London
City Museum and Art Gallery, Bristol
Fitzwilliam Museum, Cambridge
Royal Scottish Museum, Edinburgh
Victoria and Albert Museum, London

Holland

Gemeentelijk Museum 'het Princessehof', Leeuwarden
Rijksmuseum (Aziatische Kunst), Amsterdam
Rijksmuseum voor Volkenkunde, Leiden

Hong Kong

Art Gallery of the Chinese University of Hong Kong
Hong Kong Museum of Art

Iran

Archaeological Museum, Teheran

Japan

Hakutsuru Art Museum, Kobe
National Museum, Tokyo
Nezu Art Museum, Kyoto

Portugal

Museu Nacional de Arte Antiga, Lisbon

Sweden

Gustaf IIIs Antikmuseum, Stockholm
Museum of Far Eastern Antiquities (Ostasiatiska Museet), Stockholm

Switzerland

Musée Ariana, Geneva
Museum Rietberg, Zurich

Taiwan

National Palace Museum, Taipei

Turkey

Topkapi Palace Museum, Istanbul

Union of Soviet Socialist Republics

State Hermitage Museum, Leningrad

United States of America

Albright-Knox Art Gallery, Buffalo, N.Y.
Art Institute of Chicago, Chicago
Art Museum, Princeton University, Princeton, N.J.

Asian Art Museum of San Francisco, San Francisco
Brooklyn Museum, New York
Cincinnati Art Museum, Cincinnati, Ohio
Cleveland Museum of Art, Cleveland, Ohio
Dallas Museum of Fine Art, Dallas
Dayton Art Institute, Dayton, Ohio
Denver Art Museum, Denver
Detroit Institute of Arts, Detroit, Mich.
Field Museum of Natural History, Chicago
Fogg Art Museum, Harvard University, Cambridge, Mass.
Freer Gallery of Art (Smithsonian Institution), Washington, D.C.
Herbert F. Johnson Museum of Art, Cornell University, Ithaca, N.Y.
Honolulu Academy of Arts, Honolulu, Hawaii
Indianapolis Museum of Art, Indianapolis, Ind.
Los Angeles County Museum of Art, Los Angeles
Metropolitan Museum of Art, New York
Minneapolis Society of Fine Arts, Minneapolis, Minn.
Minnesota Museum of Art, St. Paul, Minn.
Museum of Art and Archaeology, University of Missouri, Columbia, Mo.
Museum of Fine Arts, Boston
Newark Museum, Newark, N.J.
Philadelphia Museum of Art, Philadelphia, Penn.
Portland Art Museum, Portland, Oregon
Seattle Art Museum, Seattle, Wash.
Virginia Museum of Fine Arts, Richmond, Va.
William Rockhill Nelson Gallery-Atkins Museum of Fine Art, Kansas City
Yale University Art Gallery, New Haven, Conn.
Young Memorial Museum, San Francisco

Bibliography

Abbreviations and Periodicals:

AA	*Acta Asiatica (Toho gakkai),* The Institute of Eastern Culture, Tokyo, 1960–
ACASA	*Archives of the Chinese Art Society of America,* formerly *Archives of Asian Art,* New York, 1945–
AO	*Ars Orientalis,* Washington, D.C. and Ann Arbor, Michigan, 1954–
ArA	*Artibus Asiae,* Dresden, 1925–40; Ascona, Switzerland, 1947–
BMFEA	*Bulletin of the Museum of Far Eastern Antiquities,* Stockholm, 1929–
Publ. EFEO	*Publications de l'Ecole française d'Extrême-Orient,* Paris and Saigon, 1903–

Chinese Periodicals:

Kaogu, Peking
Kaogu Xuebao [Chinese Journal of Archaeology], Peking, 1951–
Wenwu, Peking, 1950–

Akiyama, T.; Ando, K.; Matsubara, S.; Okazaki, T. and Sekino, T. *Arts of China: I Neolithic Cultures to the T'ang Dynasty: Recent Discoveries.* (Coordinated by Mary Tregear) Tokyo, 1968; reprinted, 1978.

Bagley, R. W. 'P'an-lung-ch'eng: A Shang City in Hupei' in *ArA* XXXIX, nos. 3, 4 (Ascona, 1977): 165–219.
Barnard, N. *Bronze Casting and Bronze Alloys in Ancient China.* Monumenta Serica Monograph XIV. Canberra and Nagoya, 1961.
Beurdeley, Michel. *The Chinese Collector through the Centuries: From the Han to the 20th Century.* (Trs. by Diana Imber) Vermont and Tokyo, 1966.
Bobot, M. T. *L'Art chinois.* Paris, 1973.
Brinker, H. *Bronzen aus dem alten China.* Exhibition catalogue: Museum Rietberg, Zurich, 1975.
Bussagli, M. *Chinese Bronzes.* [Trs. from Italian] Feltham, 1969.

Ch'en, K. *Buddhism in China: A Historical Survey.* Princeton, N. J., 1964.
Chen Mengjia. 'Xi Zhou tongqi duandai (Dating Western Zhou Bronzes)' in *Kaogu Xuebao* (Peking, 1955-6): 9–11.
Cheng Te-k'un. *Archaeology in China III: Chou China.* Cambridge, 1963.
—.*Archaeology in China II: Shang China.* Cambridge, 1960.
Ch'u Te-yi. *Bronzes antiques de la Chine appartenant à C. T. Loo et Cie.* Paris and Brussels, 1924.

Deydier, C. 'Les *Jiaguwen:* Essai bibliographique et synthèse des études' in *Publ. EFEO* CVI (Paris, 1976).
Ding Shan. *Jiaguwen suoxian shi zu ji qi zhidu.* Peking, 1956.
Dong Zuobin. *Jiaguxue wushinian* (Fifty Years of Studies in Oracle Inscriptions). [Partially translated into English] Taipei, 1955; Tokyo, 1964.

Elisseeff, D. and V. *La civilisation de la Chine classique.* Paris, 1979.
Elisseeff, V. *Bronzes archaiques chinois au Musée Cernuschi.* Paris, 1977.
Erdberg-Consten, E. von. 'Chinese Bronzes from the Collection of Chester Dale and Dolly Carter' in *ArA Supplementum XL* (Ascona, 1978).
Eskenazi, G. *Ancient Chinese Bronze Vessels, Gilt Bronzes and Sculptures: Two Private Collections, One Formerly Part of the Minkenhof Collection.* Exhibition catalogue. London, 1977.
—.*Ancient Chinese Bronzes from the Stoclet and Wessen Collections.* London, 1975.

Fairbank, W. 'Piece-mold Craftmanship and Shang Bronze Design' in *ACASA* XVI (New York, 1962): 9–15.
Fontein, J. and Wu Tung. *Unearthing China's Past.* Exhibition catalogue: Museum of Fine Arts, Boston, 1973.

Gernet, J. *Le Monde chinois.* Paris, 1972.
Granet, M. *La religion des chinois.* Paris, 1951.
Gutkind Bulling, A. and Drew, Isabella. 'The Dating of Chinese Mirrors' in *ACASA* XXV (New York, 1971): 36–57.

Hansford, H.S. *The Seligman Collection of Oriental Art.* Vol. 1. London, 1957.

Heusden, W. van. *Ancient Chinese Bronzes.* Tokyo, 1952.

Jenyns, R.S. and Watson, W. *Chinese Art: II The Minor Arts: Gold, Silver, Bronzes, Cloisonné, Cantonese Enamels, Lacquer, Furniture, Wood.* London and New York, 1963; 2nd ed., 1980.

Karlgren, B. *A Catalogue of the Chinese Bronzes in the Alfred F. Pillsbury Collection.* Minneapolis, 1952.

—and Wirgin, J. *Chinese Bronzes: The Natanael Wessen Collection.* Stockholm, 1969.

Keightley, D.N. *Sources of Shang History: The Oracle-bone Inscriptions of Bronze Age China.* Berkeley, Los Angeles and London, 1978.

Kelley, F.C. and Chen Mengjia [Ch'en, Meng-chia]. *Chinese Bronzes from the Buckingham Collection.* Exhibition catalogue. Chicago, 1946.

Kidder, J.E. Jr. *Early Chinese Bronzes in the City Art Museum of St. Louis.* St. Louis, 1956.

Koop, A.L. *Early Chinese Bronzes.* London, 1924. Reprint. New York, 1971.

Laufer, B. *Chinese Pottery of the Han Dynasty.* Leiden, 1909.

Lefebvre d'Argencé, R.Y. *Bronze Vessels of Ancient China in the Avery Brundage Collection.* San Francisco, 1977.

—.*Chinese, Korean and Japanese Sculpture in the Avery Brundage Collection.* New York and Tokyo, 1974.

—.*The Hans Popper Collection of Oriental Art.* New York and Tokyo, 1973.

Lefeuvre, J.A. 'Les inscriptions des Shang sur carapaces de tortue et sur os' in *T'oung Pao* LXI, nos. 1–3 (Leiden, 1975): 1–82.

Li Ji [Li Chi] *Anyang.* Seattle, 1977.

—and others. 'Studies of Fifty-three Ritual Bronzes' in *Archaeologia Sinica,* N.S. V (Academia Sinica, Nangang, Taiwan, 1972).

—and others. 'Studies of the Bronze *Ting*-cauldron' in *Archaeologia Sinica,* N.S. IV (Academia Sinica, Nangang, Taiwan, 1970).

—and others. 'Studies of the Bronze *Chia*-vessel' in *Archaeologia Sinica,* N.S. III (Academia Sinica, Nangang, Taiwan, 1968).

—and others. 'Studies of the Bronze *Chueh*-cup' in *Archaeologia Sinica,* N.S. II (Academia Sinica, Nangang, Taiwan, 1966).

—and Wan Chia-pao. 'Studies of the Bronze *Ku*-beaker' in *Archaeologia Sinica,* N.S. I (Academia Sinica, Nangang, Taiwan, 1964).

Lion-Goldschmidt, D. and Moreau-Gobard, J.C. *Chinese Art I: Bronzes, Jade, Sculpture, Ceramics.* (Trs. by Diana Imber) London and New York, 1961; rev. ed. 1980.

Liu E. *Tienyun Zanggui.* Shanghai, 1903.

Lodge, J.E.; Wenley, A.G. and J.A. Pope. *A Descriptive and Illustrated Catalogue of Chinese Bronzes Acquired during the Administration of John Ellerton Lodge.* Oriental Series, No. 3. Washington, D.C., 1946.

Loehr, M. *Ritual Vessels of Bronze Age China.* New York, 1968.

—.*Relics of Ancient China from the Collection of Dr Paul Singer.* New York, 1965.

—.*Chinese Bronze Age Weapons: The Werner Jannings Collection in the Chinese National Palace Museum, Peking.* Ann Arbor, 1956.

—.'The Bronze Styles of the Anyang Period' in *ACASA* VII (New York, 1953): 42–53.

Loo & Co. *Exhibition of Chinese Arts.* Exhibition catalogue. New York, 1941.

—.*An Exhibition of Ancient Chinese Ritual Bronzes.* Exhibition catalogue. Ann Arbor, 1940.

—.*Chinese Bronzes.* Exhibition catalogue. New York, 1939.

Maspero, H. 'Contribution à l'étude de la société chinoise, à la fin des Chang et au début des Tcheou' in *Bulletin de l'Ecole française d'Extrême-Orient* XLVI, no. 2 (Paris, 1954): 335–402.

—.'La vie privée en Chine à l'époque des Han' in *Revue des Arts Asiatiques* (Paris, 1933): 185–201.

—and E. Balazs. *Histoire et institutions de la Chine ancienne, des origines au XIIe siècle après J.-C.* Paris, 1967.

Masterworks of Chinese Bronze in the National Palace Museum. Exhibition catalogue. Taipei, 1969.

Minao, Hayashi. 'In Shū jidai no zuzo kiyo' in *AA* XXXIX (Tokyo, 1968): 1–117.

Mizuno, S. *In Shū seidōki togyoku* (Bronzes and Jades of Ancient China). [In Japanese with an English summary by J.O. Gauntlett]. Kyoto, 1959.

Munsterberg, H. *Chinese Buddhist Bronzes.* Rutland and Tokyo, 1967.

Pope, J.A.; Gattens, R.; Cahill, J. and Barnard, N. *The Freer Chinese Bronzes: I Catalogue; II Technical Studies.* 2 vols. Washington, D.C., 1967; 1969.

Rawson, J. *Ancient China: Art and Archaeology.* London, 1980.

Roku-cho no bijutsu (Fine Art of the Six Dynasties). Exhibition catalogue: Municipal Museum of Fine Art Osaka, Tokyo, 1976.

Rong Geng. *Shang Zhou yiqi tongkao* (Encyclopedia of Shang and Zhou Ritual Vessels). Yenjing Xuepao Zhuanhao, no. 17, 2 vols. Peking, 1941.

Sekino, Takeshi and others. *Arts of China: Neolithic Cultures to the T'ang Dynasty.* Tokyo, 1968.

Shirakawa Shizuka. *Kimbunshū.* 4 vols. Tokyo, 1965.

Sickman, L. and A.C. Soper. *The Art and Architecture of China.* Rev. ed. Harmondsworth, 1971.

Sima Qian. *Shiji* (Historical Records). 10 vols. Peking, 1959.

Sirén, O. *History of Early Chinese Art.* 4 vols. London, 1929–30.

Soper, A.C. *Chinese, Korean and Japanese Bronzes: A Catalogue of the Auriti Collection Donated to Ismeo and Preserved in the Museo Nazionale d'Arte Orientale in Rome.* Exhibition catalogue. Rome, 1966.

—.'Early, Middle and Late Shang: A Note' in *ArA* XXVIII (Ascona, 1966): 5–38.

Sugimura, Y. *The Ancient Bronze of China* (Chugoku kodoki Idemitsu bijutsu-kan sensho). Tokyo, 1966.

Sullivan, M. *Chinese Art: Recent Discoveries.* London, 1973.

—.*An Introduction to Chinese Art.* Berkeley and Los Angeles, 1961. [Rev. ed. entitled *The Arts of China*, Berkeley and Los Angeles, 1973; 1977.]

Sumitomo, K. *Sen-oku sei-sho* (The Collection of Old Bronzes of Baron Sumitomo). Kyoto, 1934.

Suzuki, Kei and Akiyama, T. *Chugoku bijutsu* (Chinese Art in Western Collections). Tokyo, 1973.

Treasures from the Bronze Age of China. Exhibition catalogue: Metropolitan Museum of Art, New York, 1980.

Trubner, H. *The Arts of the T'ang Dynasty: A Loan Exhibition.* Exhibition catalogue: Los Angeles County Museum, Los Angeles, 1957.

Umehara, S. *Nihon shūcho Shina kodō seika* (A Selection of Ancient Chinese Bronzes in Japanese Collections). 6 vols. Osaka, 1959–64.

—.*Kanan Anyo Ibutso no kenkyu* (Studies on Relics from Anyang, Honan). Kyoto, 1941.

—.*Kanan Anyo iho.* (Selected Ancient Treasures Found at Anyang, Yin Sites). Kyoto, 1940.

—.*Ōbei shūcho Shina kodō seika* (Ancient Bronzes from Collections in Europe and America). 7 vols. Osaka, 1933.

University of Michigan Museum of Art: Chinese Buddhist Bronzes. Exhibition catalogue. Ann Arbor, 1950.

Vandermeersch, L. 'Wang Dao ou la Voie Royale' in *Publ. EFEO* CXIII (Paris, 1977).

Visser, H. *Asiatic Art in Private Collections of Holland and Belgium.* Amsterdam, 1948.

Watson, W. 'On some categories of Archaism in Chinese Bronzes' in *AO* IX (Washington, D.C. and Ann Arbor, 1973): 1–13.

—.*Early Civilization in China.* London, 1966.

—.*Handbook to the Collections of Early Chinese Antiquities in the British Museum.* London, 1963.

—.*Ancient Chinese Bronzes.* London, 1962; 2nd rev. ed., 1977.

—.*Archaeology in China.* London, 1961.

—.*China before the Han Dynasty.* London and New York, 1961.

Weber, G.W. Jr. *The Ornaments of Late Chou Bronzes.* New Brunswick, N.J., 1973.

Wen Fong (Ed.). *The Great Bronze Age of China.* New York, 1980.

White, W.C. *Bronze Culture of Ancient China.* Toronto, 1956.

—.*Tombs of Old Lo-yang.* Shanghai, 1934.

Yetts, W.P. *The Cull Chinese Bronzes.* London, 1939.

—.*The Georges Eumorfopoulos Collection: Catalogue of the Chinese and Corean Bronzes, Sculpture, Jade, etc.* 3 vols. London, 1929–32.

Young, J.J. *Art Styles of Ancient Shang from Private and Museum Collections.* New York, 1967.

Zhang Guanzhi Chang Kwang-chih. *Shang Civilization.* New Haven and London, 1980.

—.*The Archaeology of Ancient China.* New Haven and London, 1963; 3rd rev. ed., 1977.

Zheng Dekun. 'Three Dated Chinese Mirrors' in *OA* I, no. 2 (Washington, D.C. and Ann Arbor, 1948).

Zui to no bijutsu (Fine Arts of the Sui and T'ang Dynasties). Exhibition catalogue: Municipal Museum of Fine Arts, Osaka, 1976.

Photo Credits

The illustrations were supplied by the author. Christian Deydier and the publishers would like to thank all those museums, galleries, dealers and private collectors who have helped to procure the material for the illustrations – a large number are published for the first time.

The photographs which were not supplied by museums are from the following sources:
Roger Asselberghs, Brussels: Cat. 23, 33
Olaf Eckverg, Stockholm: 59
Claudio Emmer, Milan: 74
Etablissements Bulloz, Paris: 82, 98
Hans Hinz, Allschwil: 9, 22, 23, 46, 67, 146, 148, 152. Cat. 14, 37, 61, 66
Images et Reflets, Paris: 11, 34, 47, 52. Cat. 65

Jean Mazenod, Paris: 6, 150. Cat. 74
O. E. Nelson Photographer, New York: 5
Prudence Cuming Associates Ltd., London: 43, 44, 56, 81, 109, 114, 120, 132, 135, 141. Cat. 7, 10, 19, 49, 50, 62
Sotheby's, London: 125. Cat. 31
Studio Lourmel 77, Georges Routhier, Paris: 1, 2, 15, 18, 37, 38, 41, 42, 45, 57, 60, 61, 71, 77, 78, 80, 86, 94, 95, 96, 97, 101, 102, 115, 129, 130, 131, 136, 137, 139, 144, 147, 151, 153, 154, 155. Cat. 6, 8, 32, 47, 59, 67, 72, 92
Taylor and Dull Inc., New York: 4, 13, 19, 30, 33, 35, 54, 55, 68
Wilfrid Walter, London: 69
Ole Woldbye, Copenhagen: 70

Acknowledgements

My thanks are due to many people who have helped to bring this book into being. First of all to Michel Beurdeley who introduced me to the publisher, to Jean-Claude Moreau-Gobard who not only opened the door to many private collections and assisted me in the choice of illustrations, but also agreed to revise and correct my manuscript. He also worked with me continuously for several months and greatly helped to increase my knowledge of Chinese art.

The Reverend Father J.A. Lefeuvre and Professor Qu Wanli acted as mediators with the Historical Institute at Taipei, from whom many of the photographs reproduced in this book were obtained.

Furthermore, I owe thanks to my teachers Léon Vandermeersch, Vadime Elisseeff and the Reverend Father Lefeuvre, who directed my studies. To their kindness and their continuing efforts I owe my knowledge of Chinese archaeology.

I express my gratitude also to many museum curators who have never failed to make me welcome and have advised me in the choice of photographic material; chief among them are Edith Dittrich of the Museum für Ostasiatische Kunst, Cologne; Peter Thiele of the Museum für Völkerkunde, Berlin, R.Y. Lefebvre d'Argencé of the Asian Art Museum, San Francisco, Chantal Kozyreff of the Musées Royaux d'Art et d'Histoire, Brussels, and the curators of the Musée de Mariemont, Belgium, the British Museum and the Victoria and Albert Museum, London, the Rijksmuseum, Amsterdam, the Royal Ontario Museum, Toronto, the Museum of Far Eastern Antiquities, Stockholm. Nor do I forget the curators of all the other museums, whose names are too numerous to mention.

I am grateful too to the collectors and art-dealers who have allowed me to photograph pieces in their possession, in particular Maurice Bérard, Alan and Simone Hartman, J.T. Tai, Giuseppe Eskenazi and others who wish to remain anonymous.

Finally I cannot forget my wife Catherine, who contributed so much to bringing this ambitious project to its conclusion.

Christian Deydier

Index

Numbers in italics refer to plate numbers.

This book was printed in September 1980 by
Hertig + Co. AG Biel, who are also responsible for the
setting.
The photolithography in colour was furnished by Vaccari
Zincografica, Modena and in black and white by Atesa-
Argraf S.A., Geneva.
The binding is the work of Burkhardt AG, Zurich.
Editorial: B. Perroud-Benson, H. von Gemmingen
Design: Franz Stadelmann
Production: Ronald Sautebin and Franz Stadelmann

Printed in Switzerland